How to use this book

This book has been designed to cover the new AQA GCSE Science curriculum in an exciting and engaging manner and is divided into six units: B1a, B1b, C1a, C1b, P1a, P1b.

The book starts with two double page spreads focussing on How Science Works which shows you how scientists investigate scientific issues, including those in our everyday life.

Each unit in the book is broken down into separate sections, e.g. for B1b, sections 3 – 4. Each section is introduced by a double page introductory spread which raises questions about what is covered in the section, acts as an introduction to the section and includes a box encouraging you to think about what you are going to learn in the section.

There are then double page content spreads which cover what you need to learn but which also cover How Science Works and the procedures you need to be familiar with to enable you to produce your internally-assessed work: the Practical Skills Assessment and the Investigative Skills Assignment.

Each section ends with an 'ideas, evidence and issues' spread which either focuses on data interpreting and evidence or on evaluating the role of science in society and the issues that concern us all in life.

Throughout the content pages there are in-text questions to test your understanding of what you have just learnt and to further your appreciation of how science can be used and what the issues are surrounding the development of science and technology.

At the end of each unit there are two double page spreads of questions to test your knowledge and understanding of the unit. They will also prepare you for the kind of questions you will meet in exams.

The words displayed in bold in the text also appear in the glossary at the end of the book with a definition.

Contents

Science Uncovered

AQA
Science
for GCSE
FOUNDATION

Ben Clyde
Bev Cox
Keith Hirst
Mike Hiscock
Martin Stirrup

Series Editor: **Keith Hirst**

www.heinemann.co.uk
✓ Free online support
✓ Useful weblinks
✓ 24 hour online ordering

01865 888058

Heinemann Educational Publishers
Halley Court, Jordan Hill, Oxford OX2 8EJ
Part of Harcourt Education

Heinemann is the registered trademark of Harcourt Education Limited

© Harcourt Education Limited 2006

First published 2006

10 09 08 07 06
10 9 8 7 6 5 4 3 2 1

10-digit ISBN: 0 435 58601 7
13-digit ISBN: 978 0 435 58601 0

Edited by Lesley Montford
Designed by Lorraine Inglis
Typeset by Ken Vail Graphic Design

Original illustrations © Harcourt Education Limited, 2006

Illustrated by Beehive Illustration (Martin Sanders, Mark Turner), Nick Hawken, NB Illustration (Ben Hasler, Ruth Thomlevold), Sylvie Poggio Artists Agency (Rory Walker).

Printed by Bath Colourbooks

Cover photo: © Superstock

Picture research by Zooid Pictures Ltd

Index compiled by Indexing Specialists (UK) Ltd

Acknowledgements
The authors and publisher would like to thank the following individuals and organisations for permission to reproduce photographs:

Keith/Custom Medical Stock Photo/Science Photo Library p iv R; Peter Gould/Harcourt Education p iv L; WowWee Toys/WowWee Group Inc. p 2 TR; GettyImages/PhotoDisc p 2 TL; Corbis pp 2 BL, 2 BR; Samuel Ashfield/Science Photo Library p 3 L; Burke/Triolo Productions/Foodpix/Photolibrary.com p 3 R; Robert Houser/Index Stock Imagery/Photolibrary.com p 4; Stephen Morton/Empics p 6; Corbis p 8; Saturn Stills/Science Photo Library p 9 T; Patrick Bennett/Corbis UK Ltd. p 9 B; Pool/Reuters/Corbis UK Ltd. p 10; Aflo Foto Agency/Photolibrary.com p 14; Francoise de Mulder/Corbis UK Ltd. p 15; GettyImages/PhotoDisc p 17; Ed Holub/Nonstock Inc./Photolibrary.com p 20; Ian Hooton/Science Photo Library p 21; GettyImages/PhotoDisc p 22; Michael Donne/Science Photo Library p 23 R; Barry Lewis/Corbis UK Ltd. p 23 L; Boulay/Bsip/Photolibrary.com p 24; Owen Franken/Corbis UK Ltd. p 27; Nick Sinclair/Science Photo Library p 28; Plainpicture/Photolibrary.com p 29; CNRI/Science Photo Library p 30 T; Dr P. Marazzi/Science Photo Library p 30 B; Zooid Pictures p 32; Ian Hooton/Science Photo Library p 34; John Beatty/Science Photo Library p 42 T; David Tipling/Photolibrary.com p 42 B; Nick Cobbing/David Hoffman Photo Library/Alamy p 43 T; C. N. R. I./Phototake Inc/Photolibrary.com p 43 B; Rob Nunnington/Photolibrary.com p 44; Photolibrary.com p 44; W Wisniewski/Picture Press/Photolibrary.com p 44; Getty Images/PhotoDisc p 46 T; David Tipling/Photolibrary.com p 46 B; Getty Images/PhotoDisc p 47 TR; E. R. Degginger/Science Photo Library p 47 BR; Daniel Cox/Photolibrary.com p 47 TL; Corbis p 47 BL; Getty Images/PhotoDisc p 48 TL; Getty Images p 48 TR; Eyal Bartov/Photolibrary.com p 48 B; Macdonald Dennis/Index Stock Imagery/Photolibrary.com p 49 TR; Louis Quitt/Science Photo Library p 49 TL; Michael Fogden/Photolibrary.com p 49 M; Rod Planck/Science Photo Library p 49 BR; Getty Images/PhotoDisc p 49 BL; D. C. Robinson/Science Photo Library p 50; Mark Fuller/Leslie Garland Picture Library/Alamy p 51; Mark Burnett/Science Photo Library p 52; Dr Jeremy Burgess/Science Photo Library p 53; David M Dennis/Photolibrary.com p 54; Samuel Ashfield/Science Photo Library p 56; GettyImages/PhotoDisc p 57; Corbis p 58; Kathie Atkinson/Photolibrary.com p 60; Robin Bush/Photolibrary.com p 61; David Fox/Photolibrary.com p 63; Clive Bromhall/Photolibrary.com p 64; John Reader/Science Photo Library p 65; Getty Images/Photodisc p 66 L; NASA Goddard Space Flight Center (NASA-GSFC) p 66 R; Dobson Agency/Rex Features p 67 T; Robert Brook/Science Photo Library p 67 B; BSIP Krassovsky/Science Photo Library p 68 T; Phil Schermeister/Corbis UK Ltd. p 68 M; Reuters/Corbis UK Ltd. p 68 B; Corbis p 69 B; www. photos.com p 69 T; Chris Rogers/Index Stock Imagery/Photolibrary.com p 70; Frederica Georgia/Science Photo Library p 71; Corbis p 73; Wei Yao/Panorama Stock Photo Co. Ltd/Photolibrary.com p 74 T; Reuters/Corbis UK Ltd. p 74 B; Victor De Schwanberg/Science Photo Library p 75; Harcourt Education p 76; Tim Ayers/Alamy p 82 TR; Getty Images/PhotoDisc p 82 BR; Carol Dixon/Alamy p 82 L; Getty Images/PhotoDisc p 83; Leslie Garland Picture Library/Alamy p 86 T; Chris Howes/Wild Places Photography/Alamy p 86 B; Adrian Muttitt/Alamy p 89 T; Maximilian Stock Ltd/Science Photo Library p 89 B; Nick Gregory/Alamy p 90 L; Martin Stirrup p 90 R; Corbis p 91; Tom Thompson/Images.com/Photolibrary.com

p 92; Dr. Jeremy Burgess/Science Photo Library p 93 R; Philip Wegener-Kantor/Index Stock Imagery/Photolibrary.com p 93 L; Biophoto Associates/Science Photo Library p 94 TR; Vaughan Fleming/Science Photo Library p 94 TL; Charles Bowman/Photolibrary.com p 94 B; Corbis p 95; Getty Images/Brand X Pictures p 96; Getty Images/PhotoDisc p 97; Ifa-Bilderteam Gmbh/Photolibrary.com p 99; Tony Waltham/Robert Harding Picture Library Ltd/Photolibrary.com p 100 B; Charles D. Winters/Science Photo Library p 100 T; Michael Barnett/Science Photo Library p 101; Getty Images/PhotoDisc p 102 T; Harmon Maurice/Images.com/Photolibrary.com p 102 B; Science Photo Library/Science Photo Library p 103 T; Paul Seheult/Eye Ubiquitous/Corbis UK Ltd. p 103 B; NASA/Photolibrary.com p 104; Joel Stettenheim/Corbis UK Ltd. p 105; Alan Novelli/Alamy p 106 L; Peter Turnley/Corbis UK Ltd. p 106 R; Damien Lovegrove/Science Photo Library p 107; Paul Glendell/Alamy p 113; Getty Images/PhotoDisc pp 115, 116 T; Martin Bond/Science Photo Library p 116 B; Prof. David Hall/Science Photo Library p 117; Bubbles Photolibrary/Alamy p 122; Sherman Ken/Images.com/Photolibrary.com p 123; Colin Erricson/Photolibrary.com p 126; Galloway Ewing/Index Stock Imagery/Photolibrary.com p 128 L; Photononstop/Photolibrary.com p 128 R; Plainpicture/Photolibrary.com p 129; AJ Photo/Science Photo Library p 131; Foodpix/Photolibrary.com p 132 T; Chris Jones/Photolibrary.com (Australia)/Photolibrary.com p 132 B; Getty Images/PhotoDisc p 134 B; Sheffer Visual Israel/Photolibrary.com p 134 T; Gerry Johansson/Bildhuset Ab/Photolibrary.com p 135 T; James King-Holmes/Science Photo Library p 135 B; IPS Photo Index/Ips Co Ltd/Photolibrary.com p 136 L; Frank Wieder/Photolibrary.com p 136 R; Corbis p 137 T; Zooid Pictures p 137 B; Getty Images/PhotoDisc p 138 T; Foodpix/Photolibrary.com p 138 B; Wendell Webber/Foodpix/Photolibrary.com p 140; Getty Images/PhotoDisc p 141 T; Michael Carter Photography/Photolibrary.com p 141 B; Peter Sapper/Photolibrary.com p 142; Tomas del Amo/Phototake Inc/Photolibrary.com p 143; Jon Arnold Images/Photolibrary.com p 145; Getty Images/PhotoDisc p 146; AP Photo/Eric Skitzi/Empics p 148 T; Corbis UK Ltd. p 148 B; Gali Danielle/Jon Arnold Images/Photolibrary.com p 150; Todd Haiman/Images.com/Photolibrary.com p 152; Illustrated London News p 153; Leslie Garland Picture Library/Alamy p 155 L; Kaj R. Svensson/Science Photo Library p 155 R; Getty Images/PhotoDisc p 156; Reuters/Corbis UK Ltd. p 163 T; Sally A. Morgan/Ecoscene/Corbis UK Ltd. p 163 B; Martin Palm/Bildhuset Ab/Photolibrary.com p 166 T; Richard Philpott/Zooid Pictures p 166 B; Ann Shamel/Images.com/Photolibrary.com p 167; iStockphoto p 168; Andrew Lambert Photography/Science Photo Library p 169 T; Luca Trovato/Foodpix/Photolibrary.com p 169 B; D. Phillips/Science Photo Library p 170; Reuters/Corbis UK Ltd. p 171; Martyn Chillmaid/Photolibrary.com p 172 T; Cardoso/Bsip/Photolibrary.com p 172 B; Getty Images/PhotoDisc pp 174, 175; Corbis p 176 TL; Martin Bond/Science Photo Library p 176 TR; Rick Stuhlsatz/Images.com/Photolibrary.com p 176 BL; Robert Slade/Alamy p 176 BR; Dan Sinclair/Zooid Pictures p 177 T; Jerry Mason/Science Photo Library p 177 B; Franceschi/OLY/Rex Features p 178 T; Winsett Bob/Index Stock Imagery/Photolibrary.com p 178 B; Getty Images/PhotoDisc/Doug Menuez/BU003420 p 179 T; Diaphor La Phototheque/Reso E. E. I. G/Photolibrary.com p 179 B; Ken Sherman/Images.com/Photolibrary.com p 180; Sheila Terry/Science Photo Library p 181; Erik Kulin/Nonstock Inc./Photolibrary.com p 182 T; Corbis p 182 B; Sheila Terry/Science Photo Library p 183; Sri Maiava Rusden/Pacific Stock/Photolibrary.com p 184 L; Rosenfeld Images Ltd/Science Photo Library p 184 R; Richard Packwood/Photolibrary.com p 185 T; Novosti Photo Library/Science Photo Library p 185 B; Dan Sinclair/Zooid Pictures p 186 L; Mike Powles/Photolibrary.com p 186 R; ATTAR MAHER SYGMA/Corbis UK Ltd. p 187; Andrew Lambert Photography/Science Photo Library p 188 R; Andy Crump/TDR/Who/Science Photo Library p 188 L; Chase p 189 T; Simon Fraser/Science Photo Library p 189 B; University of Dundee, www. sat. dundee. ac. uk p 196; NASA/Science Photo Library p 197; AP Photo/Eric Skitzi/Empics p 198 T; Zooid Pictures p 198 B; Corbis pp 199, 200; Getty Images/PhotoDisc p 202; Plainpicture/Photolibrary.com p 203 L; Juergen Stein/Stock4b Gmbh/Photolibrary.com p 203 R; Martyn F. Chillmaid p 204 L; Dan Sinclair/Zooid Pictures p 204 M; Anthony Cooper/Ecoscene/Corbis UK Ltd. p 204 R; Martin Bond/Science Photo Library p 206 R; Mauro Fermariello/Science Photo Library p 206 L; Tony Mcconnell/Science Photo Library p 207 TR; Carlos Dominguez/Science Photo Library p 207 TL; Antonia Reeve/Science Photo Library pp 207 B, 208 L; John T. Fowler/Alamy p 208 R; Tracy Pompe p 209; Neil Duncan/Photolibrary.com (Australia)/Photolibrary.com p 210; geogphotos/Alamy p 211; Cordelia Malloy/Science Photo Library p 216; Prof. J. Leveille/Science Photo Library p 217; AJ Photo/Science Photo Library p 219; Martyn F. Chillmaid/Science Photo Library p 219; European Southern Observatory/Science Photo Library/Science Photo Library p 222 T; Todd Haiman/Images.com/Photolibrary.com p 222 B; NASA/Index Stock Imagery/Photolibrary.com p 223 R; J. P. Harrington/K. J. Borkowski/University of Maryland/NASA p 223 L; NASA/ESA/P. Feldman (Johns Hopkins University)/H. Weaver (Johns Hopkins University Applied Physics Lab)/NASA p 224; Getty Images/PhotoDisc p 224; Martin Bond/Science Photo Library p 225; PhotoDisc/StockTrek p 228 B; Margaret Bourke-White/Time Life Pictures/Getty Images p 228 T; Science Photo Library p 229.

The authors and publisher would like to thank the following individuals and organisations for permission to reproduce copyright material:

British Heart Foundation Coronary heart disease statistics, p 13 M, p 17 BL; Islington NHS Primary Care Trust Stop Smoking Group, p 25 T; Cancer Research UK, p 28 B; NHS National Patient Safety Agency, p 30 BL; Bandolier website, p 33 B; Reprinted by kind permission of the New Internationalist, Copyright New Internationalist, p 77 TR; Andy Darvill, p 213 TR. All Crown Copyright material is reproduced with the permission of the Controller of HMSO and the Queen's Printer for Scotland.

Tel: 01865 888058   www.heinemann.co.uk

C1b Oils, Earth and Atmosphere

P1a Energy and Electricity

P1b Radiation and the Universe

How science works

▲ Not all scientists wear white coats.

Why study science?

Even if you are not going to become a scientist, it is important that you know how scientists work. Science affects our lives in many ways, and we need to know what scientists are up to, so that they cannot 'pull the wool over our eyes' or 'blind us with science'.

We need to:

- understand how scientific experiments are carried out and how information is collected.
- be able to tell the difference between facts and opinions.
- decide whether the information, or the people providing it, are biased in any way.
- make our own judgements, based on an understanding of the facts.

Scientists use technical terms, and you must learn what some of these mean. These words are printed in **bold** and explained in this book.

General principles

Many scientific investigations start with **observations**. These need to be made carefully and unbiased. They are often the basis for investigations or classification of things.

This factory is situated in a rural valley. For years it caused no problem until the management decided to change the type of fuel that was being used. Local residents then noticed that a lot of smoke was drifting down the valley, and were concerned for their health.

▲ Observing the smoke from this factory started a local investigation.

When the local residents protested, the factory managers employed a team of scientists to investigate the resident's observations. They took a lot of measurements. In any investigation, it is important to make measurements using the most suitable instruments. Different instruments measure with different levels of **precision**. A tape measure measures to the nearest 1 centimetre and a ruler to the nearest 1 millimetre. You also need to use instruments effectively to get accurate results. An accurate reading is one that is nearest to the true result.

After a few months, the scientists concluded that the factory chimney was safe.

The local residents were not happy, and decided to challenge the findings.

Designing an investigation

The residents wanted to know how the investigation had been designed. The first thing they asked was 'What did you actually measure?'

It is important to make sure that we are measuring the right thing in an investigation, otherwise the findings may not be **valid**. Valid results are ones that answer the original question. A valid result is one that can be matched by other scientists following your method. They should get the same result which validates your result. Valid results are also **reliable**. A value becomes more reliable the more times it is measured, for example taking 5 readings and working out the average makes a value more reliable.

The next thing that the residents wanted to know was 'Where did you put the measuring instruments?'

▶ The 4 points marked A to D are possible measuring points.

Questions

d Where would you have placed the measuring instruments? Give a reason for your answer.
e They also asked 'When did you take the measurements?' Why do you think this was important to know?
f They next asked how many times were the measurements repeated. Why is it important that the measurements should have been repeated?

Questions

a What observations do you think that the scientists should first make?
b List some of the things that you think the scientists ought to measure.

Factory chimney declared safe

Following a thorough investigation by a team of scientists, local residents can be assured that the chimney at Smith & Jones chipboard factory is perfectly safe.

Question

c Which of the following do you think it would have been most important to measure?
• the number of hours for which smoke was produced
• the amount of toxins (poisons) in the smoke
• the colour of the smoke
• the time of day at which most smoke was produced
Give a reason for your answer.

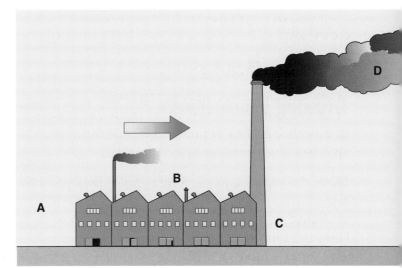

Looking at the evidence

Presenting the data

The scientists had presented their **conclusions** to the public. Conclusions must be limited to the data and evidence collected and not go beyond them. As the scientists hadn't published their evidence it wasn't easy for the residents to be sure the conclusions were valid. The credibility of **evidence** can be influenced by various things, such as if it matches current science ideas, if it is balanced or biased by the way it was collected.

The residents asked to see the **data** that the scientists had collected. Part of this is shown below.

Time of sampling	Parts per million of pollutant
8.00 a.m.	35
Noon	28
4.00 p.m.	26
8.00 p.m.	20
Midnight	15
4.00 a.m.	15
Average value over 24 hours	23

▲ Just as in a court of law, the evidence in a scientific investigation must be carefully considered.

The length of time between measurements is known as the **interval** between the measurements. Shorter intervals can give us more information: if the intervals are too big we do not know what is happening between the measurements.

Questions

a What is the interval between the measurements in the table?

b Can you think of any other information, not included in the table, which it would be important to know?

Drawing conclusions

The factory managers published this conclusion, based on the data in the table above.

Question

c Do you think that the scientists were right in coming to this conclusion? Explain your answer.

The maximum permitted amount is 30 parts per million. As the average value is well below this, the chimney may be considered safe.

Science and society

Sometimes the conclusions that people come to may be influenced by other things and are not based on the scientific evidence. For example, for many years tobacco companies were accused of misleading the public about the effects of smoking. They might have wanted to do this so that they did not lose sales of their products.

The people who lived by this factory were not prepared to believe everything they read in the newspapers. They thought about the situation carefully, and knew enough about how science works to be able to ask the right questions.

Question

d Can you think of any reasons why the company operating the factory with the smoking chimney might have been biased in coming to their conclusion?

Residents win battle over factory chimney

Local residents are delighted that their determined efforts have paid off. They have successfully challenged the results of the scientific enquiry and have been able to prove that the evidence presented was unreliable and not valid .

Your work as a scientist

These are the sorts of problems and questions that you will have to deal with in your science course. In your practical work you will need to think carefully about all of these points:

- observing
- designing an investigation
- making measurements
- presenting data
- looking for patterns and relationships
- coming to conclusions
- considering the relationship between science and society.

Finally, remember that sometimes it is difficult to collect enough evidence to be able to answer a question properly. There are also questions that cannot be answered by science alone, but need moral or social judgements to be made.

Robosapien is programmed to do 67 different tricks. He can pick up, drop and throw objects. He can also bend over and twist from side to side, so he can sit, walk and perform kung-fu moves. It took a large team of scientists many years to program *Robosapien's* computer to control these simple actions.

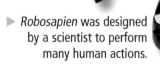

▶ *Robosapien* was designed by a scientist to perform many human actions.

◀ A toddler can walk better than *Robosapien* by the age of 18 months.

▶ This blind boy has a special stick to stop him bumping into things.

▼ This gymnast's nervous system controls her muscles. Even the best computers cannot make robots move like this.

Our nervous system also gives us information about the world around us, but we can use this information much better than a robot because we have intelligence. A robot would not know that it should not walk into a fire or off the top of a cliff! Unfortunately, some people are born with **sense organs** that do not work correctly.

Where would we be without hormones?

Many of our body processes are controlled by chemicals called **hormones**. These are produced by glands and transported around the body in the blood. They help to control the function of many organs. Understanding how hormones work has helped scientists to do many things to help our body conditions and health.

Making decisions

You make decisions every day which affect your health. For example, you choose what to eat and drink several times every day. Your decisions may be influenced by what you read in newspapers and magazines, and what you see on TV.

◀ Millions of people with diabetes have been saved by insulin treatment.

▶ Many parents would like all food advertising aimed at children to be banned.

TV adverts

In a recent survey of TV adverts it was found that 95% of the food products advertised during children's viewing time contained high levels of sugar, fat or salt. These foods, known as 'junk foods', are also advertised using toys and games. This advertising encourages young children to eat more and more unhealthy foods.

A parents' campaign group is trying to get a ban on food advertising aimed at children. They say, 'A car can't run on bad petrol, and our kids can't run on bad food.'

Many people say they need more advice on how to find out if a food is healthy. Information on food labels helps people to compare foods and make healthy choices.

Think about what you will find out in this section

How does the nervous system help us to respond to external and internal changes?	What are the pros and cons of using hormones to control human fertility?
How is the female menstrual cycle controlled?	Can we trust the information in food advertisements and on labels?
How can we evaluate the effects of food and exercise on health?	

When we exercise

Many sports clubs have vending machines with sports drinks but do these drinks actually help your body to recover faster?

When you run, your muscles produce about the same amount of heat as a radiator. To get rid of this heat your body produces sweat. Sweating is one of the ways we keep our body temperature constant.

The graph shows how the woman's body temperature changes as she 'powers-up'.

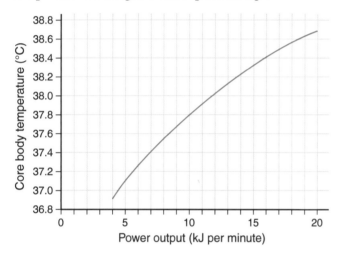

▲ Exercise produces as much heat as a radiator.

Replacing salt and sugar

When we exercise, we lose water both in sweat and in the air we breathe out. We must drink often to replace this water. If you have ever tasted sweat you will know that it is a bit salty. The saltiness is due to ions – salt consists of sodium ions and chloride ions.

When we sweat we lose a lot of water, but not quite as many ions. This leaves us with more salt in our blood than normal. If the balance of ions and water changes in our bodies, cells do not work as well. Sports drinks replace water and ions in the right proportions. Sports drinks also contain glucose. This helps to top up the glucose in an athlete's blood during a marathon. To work properly, body cells need a constant supply of glucose for their energy needs. This glucose is supplied by the blood.

Why is our body temperature kept at 37°C?

Millions of chemical reactions happen in our bodies all the time. Most of the reactions that take place in our cells are controlled by **enzymes**. Enzymes work best at a particular temperature. A healthy human body stays at 37°C. This is the temperature at which human enzymes work best. If our bodies were cooler than 37°C, the chemical reactions in our cells would be much slower.

▲ Enzymes only work properly if the temperature is just right.

Balancing the water budget

To stay healthy the body needs to balance the gain and loss of both water and ions.

The kidneys regulate the amount of water and ions in the body. They do this by producing fluid called urine. Urine contains water and salts that the body does not need.

▶ This water budget is balanced.

water in   water out

Question

c *This table shows how much water is lost from the body on a cool day. Copy the table and fill in the right-hand column to show whether the volume of water lost on a hot day would be the same, more or less. Give a reason for each of your answers.*

	Water loss in cm³	
	Cool day	Hot day
breath	400	
sweat	500	
urine	1500	

Key point

● It is important to control the internal conditions of the body. These include water content, ion content, blood sugar levels and temperature.

Detectors and sensors

Electric chip pans do not usually catch fire because they switch off automatically when the fat gets too hot. To do this they have a heat detector. We have detectors in our bodies to detect danger such as heat. These detectors have specialised cells called **receptors**. A change in the environment that can be detected by a receptor is called a **stimulus**. Our sense organs contain many receptors. The receptors that are sensitive to changes in position are found in the ears. These receptors help us to keep our balance.

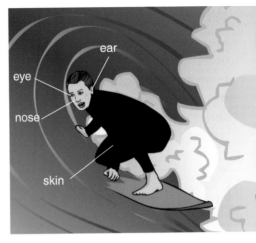

▲ Information from receptors helps this surfer to stay upright.

Question

a Which of the sense organ(s) in the diagram contains mainly receptors sensitive to: (i) sound (ii) chemicals (iii) light (iv) touch and pressure?

The girl who feels no pain

Five-year-old Ashlyn Blocker can't feel pain. She is fearless. She needs to be checked regularly for scrapes and bruises because she wouldn't know if she had cut herself or broken something.

Why can't Ashlyn feel pain, even though she has pain receptors?

Receptors only detect stimuli. The information from the receptors needs to be passed to the brain in order for the body to react. Information is carried around the body by specialised cells called **neurones**. Neurones are often found in groups called **nerves**. Information does not get to Ashlyn's brain because some of her nerves have not developed properly.

◀ The girl who feels no pain.

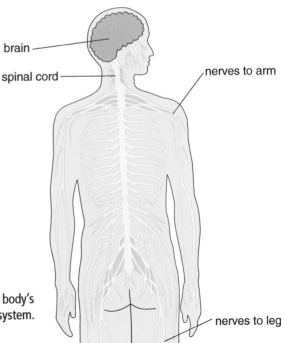

▶ The body's control system.

The nervous system

The brain and spinal cord make up the **central nervous system**. They collect information from receptors, make sense of it and then send information to the organs that need to respond. For example, if the temperature receptors in your fingers send a message to the spinal cord that they have touched something too hot, the spinal cord makes your arm muscles contract to move your hand away. The muscle is known as the **effector**.

Reflex actions

This is an example of a **reflex action**. Three types of neurones are involved in this reflex action: sensory, motor and relay. Sensory neurones carry information to the central nervous system. **Motor neurones** carry information away from the central nervous system. **Relay neurones** carry information within the central nervous system.

▶ How your nervous system stops you from getting burned.

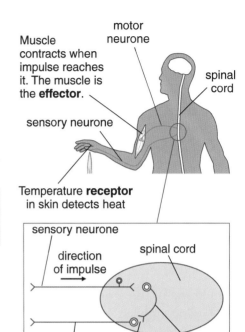

Muscle contracts when impulse reaches it. The muscle is the **effector**.

motor neurone

spinal cord

sensory neurone

Temperature **receptor** in skin detects heat

sensory neurone

direction of impulse

spinal cord

motor neurone

relay neurone

synapse

Question

b Which neurone carries information: (i) from the receptor to the spinal cord (ii) from the spinal cord to the muscle?

Information is carried along neurones as electrical **impulses**. Between each pair of neurones there is a junction called a **synapse**. The electrical impulse cannot pass across this gap. Instead, information is carried across the synapse by chemicals that are released by one neurone and detected by the next.

Not all our reflexes depend on muscle movement. Other reflexes help the body to work properly. When you smell food that you like, your mouth will begin to water. In this case it is not a muscle that brings about the reaction, but the salivary glands in the mouth. This is the mouth-watering reflex.

Two students investigated how quickly the nervous system works. The diagram shows how they did it.

The boy dropped a ruler and the girl caught it. The table shows their results.

Attempt	Distance ruler fell (cm)
1	28
2	26
3	24
4	24
5	25

Key points

- The central nervous system coordinates the body's reactions to stimuli. Receptors detect stimuli and send information along neurones to the central nervous system, which makes muscles or effectors react.
- Between each pair of neurones there is a synapse.
- Sensory neurones carry information to the central nervous system. Motor neurones carry information away from the central nervous system. Relay neurones carry information within the central nervous system.

Question

c (i) What control variables should the students have used?
(ii) In what ways could the students have improved the reliability of their results?

Getting pregnant

Many couples have been helped to have children by fertility drugs. One problem with fertility drugs is that the woman may have twins, triplets or even more babies. To understand why this happens, we have to study the hormones that control reproduction.

Getting the timing right

Once a woman reaches puberty, one of her ovaries releases a mature egg every 28 days or so. If it is fertilised, the egg cell divides to form an embryo, which attaches to the lining of the womb.

The lining of the womb needs to be ready to receive a growing embryo. At the same time that an egg is developing inside an ovary, the lining of the womb becomes thicker. This is to supply the developing embryo with food and oxygen. If the egg is not fertilised, the lining of the womb breaks down, causing bleeding. The monthly cycle of changes that take place in the ovary and the womb is called the **menstrual cycle**. The menstrual cycle is controlled by hormones. The action of the hormones involved is shown in this picture.

▲ Without fertility drugs, this couple could not have started a family.

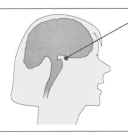

The pituitary gland produces two hormones:
• FSH, which controls when eggs in the ovaries ripen, and causes the ovaries to release the hormone oestrogen
• LH, which controls when eggs are released into the oviduct

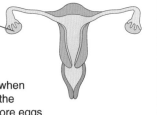

The ovary glands produce the hormone oestrogen that:
• controls when the lining of the womb thickens and when it breaks down
• sends signals to the pituitary gland when an egg is ripe and when an egg is fertilised – this stops the pituitary hormones releasing more eggs

▲ Controlling the menstrual cycle.

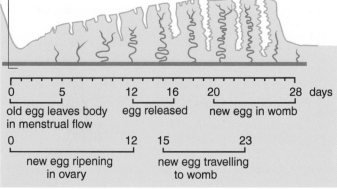

thickness of womb lining

| 0 | 5 | 12 | 16 | 20 | 28 days |

old egg leaves body in menstrual flow | egg released | new egg in womb

| 0 | 12 | 15 | 23 |

new egg ripening in ovary | new egg travelling to womb

▲ Events in the menstrual cycle.

Question

a (i) Name the hormone that causes the lining of the womb to thicken.
(ii) Name the hormone that causes an egg to be released in the middle of the menstrual cycle.

Question

b (i) On which days could an egg be released?
(ii) On which day is the lining of the womb thickest?
(iii) On which day does bleeding start?

The pill

Women can take the **contraceptive pill** to stop them becoming pregnant. The pill contains hormones that inhibit the production of **FSH**. The ovaries do not release eggs when women take the pill properly. If women remember to take the pills at the right time, this is a very reliable method of preventing pregnancy, but it can produce side effects. Some women get headaches or feel sick. In a very small number of women, the contraceptive pill can cause blood clots.

Fertility drugs

Some couples want to have children but the woman cannot become pregnant because her ovaries do not release eggs. She is infertile. This is often because her pituitary gland is not making enough FSH.

The woman can be treated by having a **fertility drug** injected into her blood. This contains the hormone FSH, which stimulates eggs to mature in the ovaries.

Unfortunately, the treatment does not always work or, sometimes, it may cause more than one egg to be released. This can result in twins, triplets, quadruplets or even more!

Test-tube babies

Some women produce normal levels of hormones but still can't get pregnant. To help these women, eggs can be taken from the ovaries. To stimulate the ovaries to release eggs, the woman is given an injection of **LH**. The eggs are then placed in a test tube and fertilised with sperm from the father. The fertilised eggs begin to divide to form embryos. The embryos are then placed in the mother's womb, where they develop naturally. Usually three **embryos** are placed in the womb in case one or more does not develop properly.

This type of fertility treatment is called **IVF – *in vitro* fertilisation**. This literally means fertilisation in a test tube, which led to the term 'test-tube baby'.

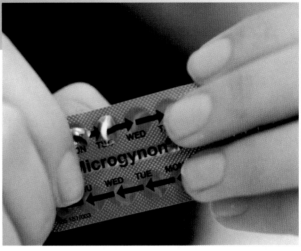

▲ A daily task for many women who don't want to get pregnant.

▲ When fertility treatment goes wrong.

▲ It's a boy!

Question

c (i) Give one side effect of using the pill.
   (ii) Give one possible problem caused by fertility drugs.
   (iii) Name the hormone used in fertility drugs.
   (iv) Name the hormone used in test-tube baby treatment.

Key points

- Hormones are chemicals that control many processes in the body, such as the menstrual cycle.
- Artificially produced hormones including oestrogen can be given to women as oral contraceptives to help control fertility.
- FSH is also called the fertility drug because it can be given to women to stimulate egg release and help them get pregnant.

The world's oldest mother

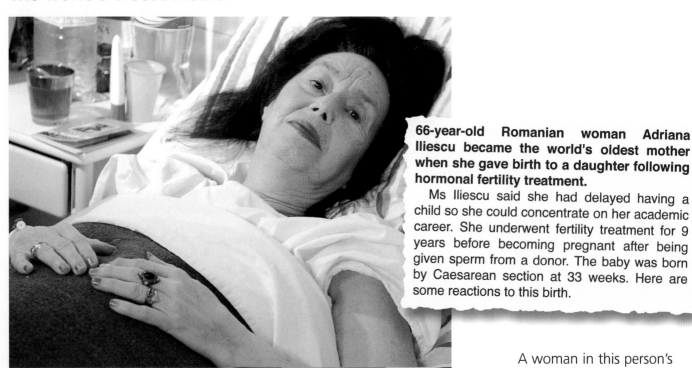

▲ The world's oldest mother.

66-year-old Romanian woman Adriana Iliescu became the world's oldest mother when she gave birth to a daughter following hormonal fertility treatment.

Ms Iliescu said she had delayed having a child so she could concentrate on her academic career. She underwent fertility treatment for 9 years before becoming pregnant after being given sperm from a donor. The baby was born by Caesarean section at 33 weeks. Here are some reactions to this birth.

A woman at grandmother age shouldn't be having children. I can see no justification for this.

The chances of her being around when the child is 18, 19 or any of the times when she will need her mother are extremely doubtful.

A woman in this person's circumstances would almost certainly not be treated in the UK. There's no upper age limit in this country but in each case you have to look at what's in the interest of the child. One of the things is age. Are you able to care for the child as they grow up?

Campaign group leader

Social worker

Spokesman from the authority that regulates the UK's IVF clinics

Question

a In groups, use the information above to debate the question 'Should women over 55 be given fertility treatment?'

The risks of the pill

Does the pill work?

Only one in one hundred women on the pill become pregnant.

The pill actually protects you from cancer of the ovary and cancer of the womb. But there is a very slight increased risk of breast cancer in people who take the pill for many years.

Does it give you cancer?

I know a girl whose aunt got blood clots after taking the pill.

Health visitor

Does the pill help in any other way?

It makes your periods more regular, and it might even help to clear up acne.

The risk of blood clots is 5 per 100 000 for women who are not on the pill and rises to 15 per 100 000 for women on the pill. But for women who are pregnant, the risk of a blood clot is 60 per 100 000.

Question

b *Make a summary of the benefits and problems linked with using the pill.*

Evaluating sports drinks

Sports drink	Carbohydrate (g/l)	Sodium ions (mmol/l)	Chloride ions (mmol/l)
Coca Cola	105	3	1
Dioralyte	16	60	60
Gatorade	62	23	14
Isostar	73	24	12
Lucozade	180	0	0
Lucozade Sport	64	23	14

Question

c *The table above gives the composition of some sports drinks.*
(i) Which drink would give you most energy?
(ii) Which drink would restore the ions lost through sweat the quickest?
(iii) Which drink would be worst for a marathon runner? Explain the reason for your answer.

Key points

- There are benefits and problems arising from the use of hormones to control fertility which need to be considered carefully before use.
- We can evaluate the claims of manufacturers for their sports drinks by considering what the body needs and whether or not the drink supplies it.

Eating habits

This article shows how worried doctors are about the effects of poor eating habits and lack of exercise among young people.

The table below shows some of the results of two recent surveys about eating habits.

FAT EPIDEMIC WILL CUT LIFE EXPECTANCY

The childhood obesity epidemic caused by poor nutrition and lack of exercise is creating a health crisis. The average life expectancy is expected to drop for the first time in more than a century.

The chairman of the Food Standards Agency said that obesity was a 'ticking time bomb' and was one of the most serious issues facing the nation.

Food	Average consumption in 1992	Average consumption in 2000
fresh vegetables	161 g/person/day	180 g/person/day
fresh fruit	132 g/person/day	177 g/person/day
soft drinks containing sugar	720 ml/person/week	1284 ml/person/week
chocolate	35 g/person/week	124 g/person/week

Questions

a How much more vegetables were eaten in 2000 than in 1992?

b Food experts say that eating snacks between meals has increased. What evidence in the table supports this view?

You are what you eat

The food you eat provides the energy you need to stay alive and be active. Your food also provides the proteins, vitamins and minerals your body uses to grow and replace damaged cells and tissues. A diet that gives all this is a balanced diet.

▶ These are the proportions of different foods that make up a healthy balanced diet. Does this match your diet?

Energy in and energy out

You need to eat enough food each day to provide the energy your body needs. If you eat more food than your body needs, your body stores the extra food as fat and you put on weight. The amount of energy a person needs depends on how active they are.

Too little food
Energy used is greater than energy intake = too thin

Question

c What happens if you don't eat enough food to give you the energy your body needs?

Burning food to give energy

Your body gets energy from food by respiration. The rate at which chemical reactions are carried out in the cells of your body to give energy is called the **metabolic rate**. Your metabolic rate is affected by:

- the amount of exercise you do
- the proportion of muscle to fat in your body
- the genes you have inherited.

Too much food
Energy intake is greater than energy used = too fat

Lifestyle trends

You use more energy to maintain your temperature in cold weather and less energy in warm weather.

Eating too much food and not taking enough exercise can lead to people becoming overweight. These days, many people have very inactive lifestyles. Watching TV, playing computer games and getting a lift in the car all use much less energy than walking to school or playing sport.

People who exercise regularly are usually fitter and healthier than people who take little exercise. People like this are generally more aware of what they eat and try to eat healthily.

The graph shows the results of a survey of over 6000 young people from all over the UK.

▲ Percentage of young people taking part in at least 60 minutes of physical activity per day (national survey 2002).

Key points

- A balanced diet supplies the correct amount of energy and nutrients for a healthy body.
- The metabolic rate is the rate at which the chemical reactions in the body happen. This depends on how much you exercise, the proportion of muscle to fat in your body and inherited factors.
- People who exercise regularly are usually fitter than those who don't.

Questions

d At what age is activity among girls at its lowest level?

e Doctors recommend that young people carry out at least 60 minutes of moderate exercise every day. What percentage of 15-year-old boys achieves this recommendation?

f Why were large numbers of people used in the survey?

Health problems

In developed countries such as the UK, people are eating too much food and taking too little exercise, causing health problems.

The World Health Organization recently carried out a survey of the diets of children in 35 countries. They found that the diets of children in the UK were among the worst.

In the long term people who are overweight are more likely to develop health problems such as arthritis, diabetes, high blood pressure and heart disease. You can find out more about blood pressure and heart disease on page 16.

Arthritis and being overweight

The bones in your joints move easily because they are covered by cartilage. The cartilage may get worn away with use so the bones scrape against each other. This causes the joints to become stiff and painful. This is called **arthritis**.

People who are overweight have more weight pressing down on their knee and hip joints. The cartilage in their joints wears away more quickly, causing arthritis to develop.

> ### Question
>
> **a** Why are overweight people more likely to develop arthritis?

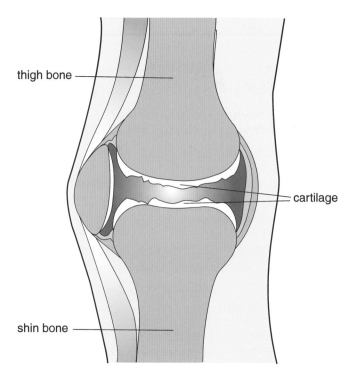

▶ Knee joint showing damage to cartilage, which causes arthritis.

thigh bone

cartilage

shin bone

Diabetes and controlling blood glucose

Diabetes is an illness in which the body cannot control the amount of glucose in the blood. Type 2 diabetes usually develops in people over the age of 40. Overweight people are more likely to develop this form of diabetes. About 80% of people with this condition are overweight when they are first diagnosed.

Sugary drinks – are they good for you?

Sugary drinks are drinks with added sugar. To find out if sugary drinks affect people's health, scientists carried out a survey involving 50 000 American women. They analysed questionnaires the women completed about their diet and health.

The scientists' findings were:

- Women who drank more than one sugary drink per day put on more weight than women who drank less than one sugary drink per week.
- Women whose consumption of sugary drinks was high were almost twice as likely to develop type 2 diabetes compared to women whose consumption was low.

Questions

b Why were the women in the survey asked to complete questionnaires?
c Which group had the highest risk of developing diabetes?

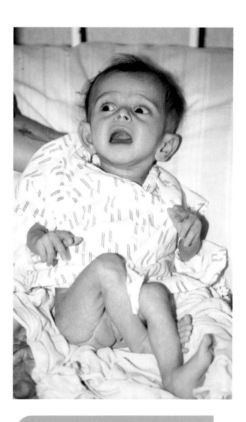

Eating too little

3 Million Facing Death in Famine

More than 3 million people, including almost a million children, are facing death from starvation in parts of Africa.

Families are feeding their children with grass and leaves to try to keep them alive.

In some developing countries many people cannot get enough food. People become malnourished and suffer major health problems if their diet doesn't provide the energy and nutrients their body needs. As well as becoming very weak people who are starved of food are more likely to:

- get infected with disease because their resistance is low
- develop **deficiency disease** because their food does not contain enough protein, vitamins and minerals
- have irregular periods (in women).

Key points

- In the developed world, eating too much and exercising too little are leading to obesity and related diseases, including arthritis and diabetes.
- In the developing world, famine leads to infection due to reduced resistance to disease and to irregular periods in women.

Heart disease

It is important to have a healthy heart but there are several types of heart disease that can prevent people being healthy. By understanding how heart disease is caused, scientists can give people advice on how to avoid it. In a healthy heart, the walls of the arteries are smooth so that blood flows easily. Too much food and too little exercise can cause fat deposits to build up inside the arteries. These deposits reduce the diameter of the artery, making it harder for blood to flow. This means the heart muscle receives less oxygen and is weaker. It may even lead to a heart attack.

> **Question**
>
> **a** Why does less oxygen get to heart muscle when people develop heart disease?

Fighting cholesterol

Cholesterol is a fatty substance that is mainly made in your liver. Your liver makes cholesterol from the saturated fats in your food. Too much cholesterol in the blood can increase your risk of getting heart disease.

Cholesterol is transported in the bloodstream attached to proteins. The combination of protein and cholesterol is called **lipoprotein**. There are two types of lipoprotein:

- low-density lipoproteins (LDLs), which carry cholesterol from your liver to your cells
- high-density lipoproteins (HDLs), which carry the extra cholesterol that your cells don't need back to your liver.

LDLs are 'bad' cholesterol and can cause heart disease. HDLs are 'good' cholesterol. The balance of LDLs and HDLs is very important in keeping your heart healthy.

Blocking blood vessels

Too much cholesterol causes fat to build up on artery walls, causing heart disease. The amount of cholesterol your liver produces depends on a combination of what you eat and inherited characteristics. The amount and type of fat you eat can change the amount of cholesterol in your blood. Saturated fats, found in animal fat, increase blood cholesterol levels. Unsaturated fats (called mono- and polyunsaturated fats), found in fish and vegetable oils, may help to reduce blood cholesterol levels.

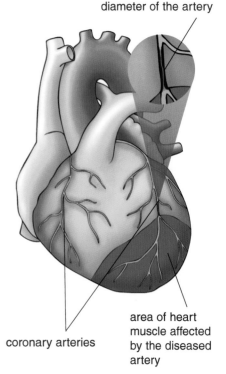

fat builds up in the artery walls, reducing the diameter of the artery

coronary arteries

area of heart muscle affected by the diseased artery

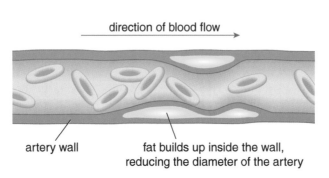

direction of blood flow

artery wall

fat builds up inside the wall, reducing the diameter of the artery

Bread:

Nutrition Facts
Per 2 slices (64g)

Amount	% RDA
Calories 140	
Fat 1.5g	2%
Cholesterol 0mg	0%
Sodium 290mg	12%
Carbohydrate 26g	9%
Protein 5g	12%

Iceberg Lettuce:

Nutrition Facts
Per Serving

Amount	% RDA
Calories 15	
Fat 0g	0%
Cholesterol 0mg	0%
Sodium 10mg	0%
Carbohydrate 3g	1%
Protein 1g	1%

Banana Cream Pie:

Nutrition Facts
Per 1/8th of pie

Amount	% RDA
Calories 470	46%
Fat 30g	89%
Cholesterol 30mg	10%
Sodium 360mg	15%
Carbohydrate 48g	16%
Protein 4g	9%

Milk:

Nutrition Facts
Per cup

Amount	% RDA
Calories 120	
Fat 5g	8%
Cholesterol 20mg	15%
Sodium 120mg	5%
Carbohydrate 11g	4%
Protein 9g	17%

Questions

b Which of the foods shown opposite is the unhealthiest? Give a reason to support your answer.

c What sort of foods would you recommend to a friend needing a low cholesterol diet?

d What sort of food would you recommend to a friend needing a low salt diet?

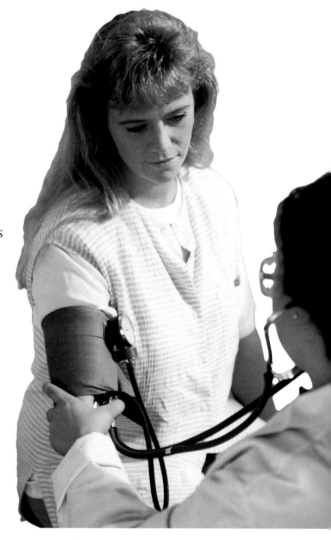

Fighting high blood pressure

People with high blood pressure are more likely to develop heart disease. Blood pressure means the pressure of the blood flowing in your arteries. A certain amount of pressure is needed to keep the blood flowing. Too much salt in your diet can lead to increased blood pressure. The amount of salt people eat has increased because many foods, such as canned soups, take-aways and ready-prepared meals, contain large amounts of salt.

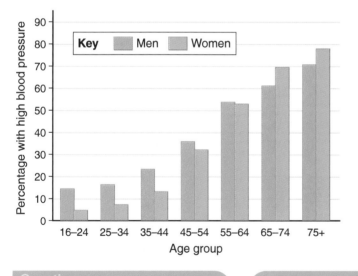

Question

e Which age group has the highest percentage of people with high blood pressure?

Key points

- High cholesterol levels in the blood increase the risk of heart disease and blocked arteries.
- There are two types of cholesterol. Low-density lipoproteins (LDLs) are bad and can cause heart disease. High-density lipoproteins (HDLs) are important for a healthy heart.
- Saturated fats increase bad cholesterol. Mono- and polyunsaturated fats reduce bad cholesterol levels.

Can you believe what you read?

The latest trends in slimming diets and slimming products appear regularly on TV and in magazines, claiming that they can help people lose weight effortlessly. Some adverts use 'evidence' and scientific jargon to try to convince people that a diet really works.

The advert on the right shows how one company tries to get people to buy their slimming pills as a way of losing weight. Does the 'proof' the advert refers to provide reliable evidence that convinces you that this product works? This advert is typical of many companies. Most of the 'evidence' is based on reports from just one or two people showing dramatic weight loss in a short time.

Health experts warn that if a product or a diet programme sounds too good to be true it probably isn't good for you and isn't true.

Health experts also say that there is only one way to lose weight – eat healthy foods, keep energy intake below energy output and aim at a sensible weight loss.

A scientific evaluation

Meal replacements are widely used as a way of losing weight. Meal replacements, such as Slim-fast, are energy-reduced products that contain added vitamins and minerals.

Three hundred people were interviewed to take part in an investigation to compare a meal-replacement diet with a conventional low-calorie diet. Sixty-six people were chosen to take part in the study. All had a similar level of health and fitness. Those chosen to take part were divided at random into two groups. One group received the meal replacement and the other group, the control group, received a low-calorie diet. Weight change was measured after 3 months and after 6 months. The results are shown in the table.

	Mean weight loss (kg)	
	Meal replacement group	Control group
after 3 months	6.0	6.6
after 6 months	9.0	9.2

Questions

a Why were a large number of people used in this investigation?
b Is using a meal-replacement diet an effective way to lose weight? Use the results to support your answer.

This shows that there was very little difference between the two groups.

Slimming programmes

People who are overweight can lose weight gradually by following a sensible slimming programme. This involves eating less energy-rich food and taking more exercise. The information on food labels can help you choose foods that are lower in fat and energy. By following a sensible slimming programme, weight will be lost gradually. Some people become so worried about being fat that they start to eat far too little. Their weight can drop dangerously low. There are health risks for people who lose too much weight too quickly.

The graphs show the changes in body mass of two young women who are trying to maintain a healthy weight. Bes controls her weight by avoiding foods high in fats and sugars, and taking regular exercise. Jo is always switching from one diet she has read about to another and often goes for days eating very little.

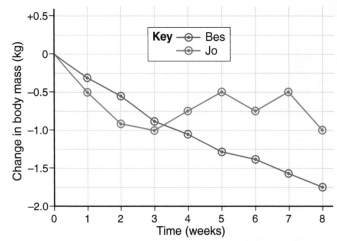

Questions

c How much weight did:
(i) Bes (ii) Jo
lose over the 8 weeks?
d Calculate the mean mass lost per week by the two women.
e Suggest reasons why Bes' methods to control her weight are healthier than Jo's.

Key points

● Data can be used to evaluate claims about slimming products and slimming programmes.
● It is necessary to distinguish between opinion based on valid and reliable evidence and opinion based on non-scientific ideas.

Drugs are chemicals that affect our body processes. Medicines are drugs designed to help us if we are ill. Other drugs are for recreational use but misusing any drugs causes problems.

Many drugs are extracted from natural substances, such as opium from poppy seeds. Natural drugs have been used by people as part of their culture for thousands of years. There were wars over opium in China in the 1800s. Alcohol has been fermented from fruit and grain since at least ancient Greek cultures.

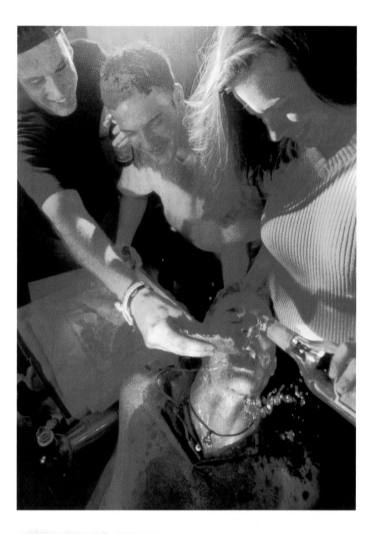

▶ Binge drinking is reported to be on the increase. Alcohol is linked to violence and anti-social behaviour.

Worrying reports

You are probably protected against measles, mumps and rubella. The MMR vaccination is given to young babies to provide protection against these three diseases. Deciding whether or not to have a young child vaccinated can be a very worrying decision for parents when they read reports saying that the MMR vaccination may harm their child. The following reports appeared in recent newspapers.

WE WISH WE'D HAD THE MMR

Cases of measles are soaring as parents reject the MMR vaccination amid fears that it may be harmful.

One parent whose daughter, Clara, almost died from the complications of measles said,

"Even now it hasn't sunk in that my daughter almost died of measles. As she lay in intensive care the doctors said it was touch and go.

I thought measles was just a harmless childhood illness. I'd had it as a girl with no problems. Now it could rob me of my daughter.

All my other children had the MMR jab. But when Clara was due for her jab she had a bad cold so I delayed the injection. That delay almost cost me my daughter."

WHY I WOULDN'T GIVE MY BABY THE MMR JAB

No one said being a parent is easy. The most difficult problem I have faced so far is the question of MMR – the measles, mumps and rubella vaccine.

Over the past six months it has become the most hotly debated subject among the mothers I know. Before my daughter was born I ignored this issue. Now she is one year old and I have to make a decision. But what shall I do?

When I see my daughter running around I feel sick at the thought that I could do her harm. All the information about the dangers of MMR has left me very confused. Only one of my friends is getting her child vaccinated with MMR. The information provided for parents doesn't convince us that the vaccine is safe. I can't take the risk of letting my child have the vaccine.

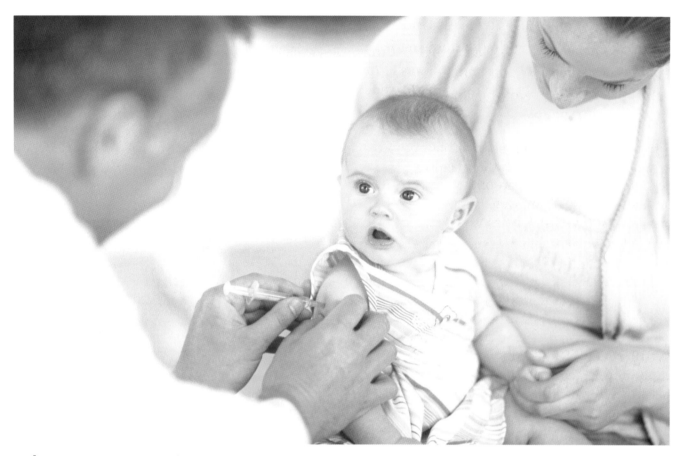

Who can you trust?

Doubts over the safety of the MMR vaccination have often been reported on TV and in newspapers. Following these reports, many parents decided not to have their child vaccinated using MMR.

The main issues about the safety of MMR are:

- nearly all the evidence gathered from several countries shows that MMR is safe
- the research of one doctor raised concerns about health risks
- some reports presented information carefully, but some reports emphasised parents' worries and fears
- recent evidence shows that the research findings about health risks were not reliable.

Think about what you will find out in this section

What are the dangers of using common drugs such as alcohol and tobacco?	Is there enough evidence to suggest a link between taking cannabis and developing mental illness?
The advantages and disadvantages of being vaccinated against diseases.	How scientists make sure that new medicines are safe to use.
How scientists make sure that their evidence is reliable.	How the treatment of disease has changed.

A drug addict at 5

There is a 5 year-old who, if not given cannabis, throws herself on the ground and screams until she gets it. Her sisters, who are aged 9 and 13, also use cannabis and drink alcohol.

Recreational drugs

Today, people take recreational drugs for many reasons. If you are feeling tired, you might drink a cup of coffee to wake you up. Sometimes people just use drugs to change their mood, for example by drinking alcohol or smoking cigarettes. People who smoke and drink need to understand the risks they are taking even though these are legal drugs. They need to look at the scientific evidence about the risks, then decide if the pleasure they get from the drug is worth it in the long run. Some drugs are so dangerous that they are illegal in Britain. Heroin is highly addictive and causes lots of health problems. Cannabis is another illegal drug. Many scientists now believe that it causes mental health problems in many users, but others are not so sure about the link.

How do people get addicted to drugs?

Some drugs are medicines designed to help you when you are ill, while other recreational ones may change the way you think or feel pain. All drugs are chemicals that affect the way your body works. They change certain chemical reactions in your body. When your body gets used to the change it may become dependent on the drug – you have become **addicted**. People can become addicted to medicines such as sleeping tablets too.

Drugs like heroin and nicotine are very addictive. People may become dependant on these drugs very quickly. Stopping taking the drug can make them very ill for a while – they suffer from **withdrawal symptoms**. For smokers trying to give up, these symptoms include:

- craving for a cigarette
- trouble sleeping
- irritability or anger
- restlessness
- trouble concentrating.

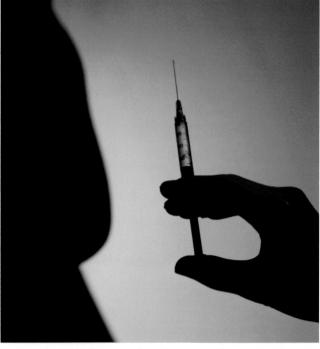

▲ Heroin addicts have to inject themselves frequently otherwise they get severe withdrawal symptoms.

Questions

a *Copy and complete the sentence.*
   Drugs change some _____ in our bodies.
b *What is meant by 'withdrawal symptom'?*

The oldest drug – alcohol

Alcohol is a drug that affects the nervous system. In moderate amounts it makes most people feel more relaxed and sociable, but even small amounts slow down the body's conscious actions so your reaction times get slower. That is why it is illegal to drive a car with just a small amount of alcohol in your blood.

Larger amounts of alcohol lead to lack of self-control. One example of this is the fights that often break out when 'lager louts' drink too much.

Party nights can be fatal. Drinking too much alcohol can cause a person to lose consciousness and even go into a coma. Every year the media report stories of young people dying from drinking too much during a celebration.

▲ A common sight in many city centres.

Drinking too much alcohol has long-term effects too. It can lead to alcohol addiction. Many alcohol addicts die early because of damage to their liver or brain.

When we add together the cost of policing binge drinking and treating the long-term effects of drinking alcohol, the overall impact of this legal drug is far greater than the impact of illegal drugs such as heroin.

LIFE-SAVING LIVER FOR YOUNG LIVERPOOL LASS

A GIRL of 20 has become the youngest person in Britain to need a new liver because of binge drinking. The girl involved told doctors at the Royal University Hospital in her home city of Liverpool that she had become a frequent binge drinker by the age of 14 and continued until she was 17. Medics found her liver was so badly damaged that she needed a new one to save her life.

▲ Breathalysers are used to deter people from drinking and driving.

Question

c *Copy and complete the sentences.*
Alcohol affects the _____ system by slowing down our _____. Drinking too much alcohol over a long period can damage our _____ and _____.

Key points

- Drugs change the chemical processes in the body so people may become addicted to them and suffer withdrawal symptoms.
- People use legal and illegal drugs recreationally. The impact of legal drugs on health is much greater than that of illegal drugs as more people use them.
- Alcohol slows down the reactions of the nervous system, helping people to relax, but too much can lead to unconsciousness or coma, and liver and brain damage.

Tobacco smoke is a mixture of hundreds of different chemicals. Many of these chemicals are dangerous to the human body. Tobacco smoke is particularly damaging to the lungs, heart and blood vessels. The chemicals in tobacco smoke cause cancer to develop in some people. These substances are known as **carcinogens**.

▶ You wouldn't buy food from a supermarket if it said it would kill you.

Underweight babies

One of the poisonous substances in tobacco smoke is **carbon monoxide**. This prevents red blood cells from carrying oxygen – so a smoker's blood carries less oxygen. This is one reason why babies born to women who smoke are lighter at birth on average than babies born to non-smokers.

The table shows the risk of getting lung cancer for smokers and non-smokers. The risk for non-smokers is given a value of 1.00. A value of 2.00 means that the risk is twice that of a non-smoker.

Smoking status		Risk of developing lung cancer
non-smoker		1.00
previous heavy smoker		2.46
current smoker	Duration of smoking in years	
	1–20	1.85
	20–39	3.74
	40+	6.02
	Number of cigarettes smoked per day	
	1–20	1.94
	20–39	3.38
	40+	4.61

▲ Smoking affects unborn babies.

Questions

a Describe, in as much detail as you can, the effect on the risk of developing lung cancer of: (i) the number of years a person smokes (ii) the number of cigarettes smoked per day.

b By how much does the risk of getting lung cancer increase when you move from smoking 10 cigarettes per day to 25 cigarettes per day?

c A person has been smoking for 40 years, then gives up. How does this affect his chances of developing lung cancer?

Ways to stop smoking

This poster was the winner in a competition to advertise a helpline for people who want to stop smoking. It was painted by a 10-year-old girl.

There are two common methods of stopping smoking:

- *Cold turkey*. This means stopping without any kind of aid. Withdrawal symptoms are very severe in the first few days but they fade away within the first two or three weeks. Most people give up smoking using this method.

- *Nicotine replacement therapy* (NRT). NRT is clinically proven to be twice as effective as the cold turkey method. NRT eases withdrawal symptoms while the smoker gets used to not smoking and the dose is gradually reduced. NRT methods include gum, skin patches and lozenges.

For more information please contact Islington Stop Smoking Service on freephone

0800 093 90 30 Islington NHS Primary Care Trust

By Isobel Cross

Questions

d What is meant by 'clinically proven'?

e What type of investigation might have been done to justify the claim that NRT is more effective than cold turkey in stopping smoking?

f The graph shows the results of an evaluation of a stop-smoking course.

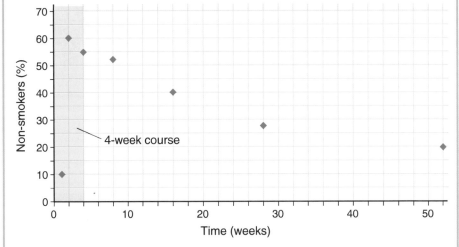

Each person on the course was given a 1-hour counselling session once per week for four weeks, together with NRT. 300 people started the course.
(i) What was the highest percentage of people who stopped smoking during the course?
(ii) How many people were still not smoking 1 year after the course began?

Key points

- Nicotine is the addictive substance in tobacco smoke. Tobacco also contains carcinogens and carbon monoxide. Carbon monoxide in the blood reduces the oxygen in the blood so pregnant women who smoke may give birth to underweight babies.

- There are a number of different ways people can be helped to give up smoking.

Developing new medicines

The treatment of disease is always being improved by the development of new drugs. Before a new drug can be used, it is put through several tests. The first tests are carried out in laboratories to find out if the drug is toxic. After this, people may be asked if they would like to be part of a **clinical trial**. Trials are carried out to find out if a new drug works better than an existing one and to make sure it is safe to use.

Heart drug could save lives

The number of people having heart attacks could be cut by at least a third by encouraging people to take drugs called **statins**. These drugs lower the amount of cholesterol in the blood. Medical experts believe that the lives of 10 000 Britons a year could be saved by increasing the number of people taking statins.

Testing a new drug

The effectiveness of statins in reducing the risk of heart attacks and other circulatory disorders was investigated by the Heart Protection Study. This study was carried out by a team of heart specialists at a leading UK hospital. The study involved 20 536 patients with heart disease. The health of these patients was monitored closely over a five-year period.

A total of 10 269 of the patients took a statin tablet daily, whilst 10 267 received a placebo every day. The placebo was a tablet that had no effect on cholesterol levels. Patients were randomly placed into the 'statin' group or the 'placebo' group, and were not told which tablets they were receiving during the trial.

At the end of the study the researchers concluded that taking statins over five years would prevent major circulation problems such as heart attacks and strokes. The study also showed that statins are very safe drugs to take.

▲ Statins are so valuable in preventing heart disease and are so safe that they can now be bought without a doctor's prescription.

Questions

b This type of study is called a randomised trial. What feature of the study was randomised?

c Why was a placebo group used in the study?

d People need to be sure that statins are safe to use and help to lower the risk of heart attacks. What features of this study ensured that the results are accurate and reliable?

An unsafe drug

Thalidomide is a drug that was first used in the UK in 1958. It was thought to be a safe drug because it had been tested and trialled. Thalidomide was given to women in the first few months of pregnancy to overcome morning sickness. Many women who took the drug gave birth to babies with limbs that weren't properly formed. The drug was banned in 1961 after it was discovered that it caused tragic birth defects.

Thalidomide is now being used again to treat leprosy. This is a severe disease that affects the skin and nerves in the hands and feet. In severe cases of leprosy the skin dies, causing hands and feet to become deformed.

▲ Leprosy is a very painful disease.

Question

e *Before a female patient with leprosy can be given thalidomide, she must first be tested to see if she is pregnant. Explain why a pregnancy test is necessary.*

Key points

- New drugs are tested to see if they are toxic and then trialled.
- Thalidomide is an example of a new drug that was tested and trialled and thought to be safe.
- Statins are new drugs being developed and tested which lower the cholesterol level in the blood, so treating and preventing heart disease.

The scientist who saved millions of lives

In the 1940s, doctors were very worried because they did not know why cases of lung cancer had risen by 50 times over just a few years.

Sir Richard Doll was the first to try to find evidence for the increase in lung cancer. In 1949 he visited 2000 people with suspected lung cancer in several different hospitals. He found that nearly all those who were later found to have the disease were heavy smokers. Most of those who were given the all-clear didn't touch tobacco.

▲ Sir Richard Doll, the scientist who discovered the link between smoking and disease.

Question

a (i) What type of investigation did Sir Richard carry out?
(ii) Why did he visit 2000 patients?
(iii) What did the results of this investigation show?

When his findings were published, Sir Richard expected people to stop smoking in their thousands. He said, 'Practically no notice was taken of it. If there was a report of it on the radio, the reporter would always be careful to mention a doctor so-and-so, who had been put up by the tobacco industry, saying the research was controversial and the results did not prove that smoking caused lung cancer'.

Sir Richard then did another investigation. He asked 40 000 doctors about their smoking habits. He then predicted which ones would be most likely to die from lung cancer. Over the next few years 37 of these doctors died from lung cancer. All of them were smokers and most were heavy smokers. Sir Richard's predictions were correct.

Question

b Suggest why the radio producers were very careful to say that the results were controversial.

Question

c Explain why this investigation provided more powerful evidence for the association between smoking and lung cancer than the first investigation.

It wasn't until 1954 that Health Minister Iain Macleod held a conference to announce that the Government accepted the association. He chain-smoked while he made the announcement.

Question

d What message was given to the public by the Health Minister smoking while he made the announcement?

The graph shows changes in smoking habits since 1950.

Question

e (i) Describe how the percentage of men who smoked changed between 1950 and 2000.
(ii) Use information from the passage and the graph to suggest an explanation for the changes.

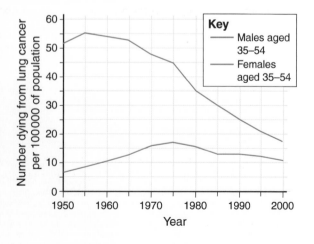

The harmless weed?

Many people smoke cannabis as a recreational drug – it helps them to 'chill out'. Most of these people think that cannabis is harmless but it may cause mental illness in some users.

The active ingredient in cannabis is a chemical called THC. Drug suppliers have bred cannabis plants that contain 20 times more THC than the wild variety. People who smoke this cannabis are much more likely to develop mental illness than those who smoke the wild variety.

▲ Many people think that smoking a joint is harmless – but is it?

Should cannabis be legalised?

Most of the heroin addicts I have to treat started on cannabis.

Most of my clients have tried cannabis, but hardly any have gone on to use heroin.

Sixty per cent of people aged 20–24 have tried cannabis, but only 1% of these people have tried hard drugs.

Health worker

Social worker

Health service spokesperson

Legalise Cannabis spokesperson

Drugs are classified into three groups: A, B and C. The table shows the maximum penalties for possessing and dealing in these drugs.

Cannabis is the most widely used drug among 11- to 19-year-olds. Twelve per cent of 11- to 15-year-olds have tried cannabis. Twenty-five per cent of those aged 16 to 19 years are using cannabis.

Drug class	Examples	Maximum penalty for possession	Maximum penalty for dealing
A	heroin, cocaine, ecstasy	7 years in prison	life in prison
B	amphetamines such as speed	5 years in prison	14 years in prison
C	cannabis, tranquillisers	2 years in prison	14 years in prison

Question

f In 2004 the government decided to downgrade cannabis from a class B drug to class C. Do you think they were right to do this? Explain the reason for your answer.

Key points

- It is important to evaluate the claims made about the effect of cannabis on health, and to look at the link between cannabis and addiction to hard drugs before legalising it.
- The link between smoking tobacco and lung cancer gradually became accepted.

Pathogens

Bacteria and **viruses** are types of microorganism. Some miroorganisms can cause disease when they get inside your body. Microorganisms that cause disease are called **pathogens**.

Bacteria are very tiny cells. In favourable conditions, like the inside of your body, bacteria reproduce very quickly. Once they are inside your body they can damage cells and produce poisons or toxins, making you feel ill.

Viruses are many times smaller than bacteria. Viruses can only reproduce inside living cells. When a virus invades a body cell it uses it to make thousands of new viruses. The new viruses then burst out of the cell and invade other body cells. This damages the cell.

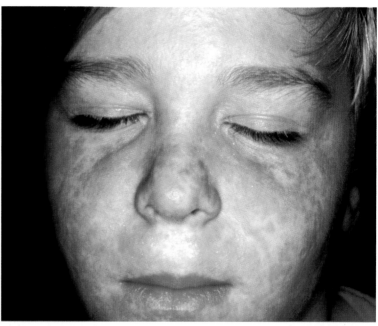

▲ Measles used to be a common childhood disease and is caused by a virus.

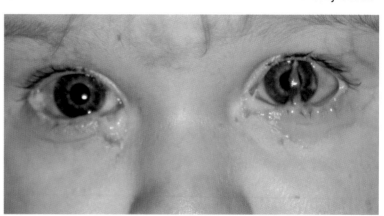

▲ This eye infection is caused by bacteria.

Spreading disease

The first person to recognise the importance of a clean environment to control the spread of disease was a Hungarian doctor called Semmelweiss. He worked in hospitals in the 1840s. At that time doctors didn't know what caused disease, so no attempt was made to stop the spread of pathogens. Semmelweiss realised that as doctors went from one patient to another they could be spreading diseases. He made all the doctors working with him wash their hands after an operation and before visiting a new patient. Deaths on the hospital wards where Semmelweiss was in charge fell from 12% to just 1%.

Cells to fight pathogens

If pathogens do get into your body you are protected by your **white blood cells**. These cells help to stop pathogens from reproducing inside your body.

One way your white blood cells protect you is by **ingesting** (or taking into the cell) pathogens. Once the pathogen has been ingested, the white blood cell releases enzymes to digest and destroy it.

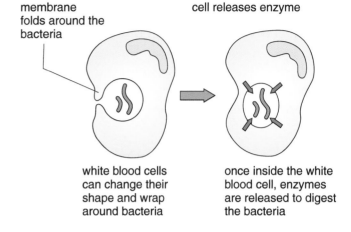

membrane folds around the bacteria

cell releases enzyme

white blood cells can change their shape and wrap around bacteria

once inside the white blood cell, enzymes are released to digest the bacteria

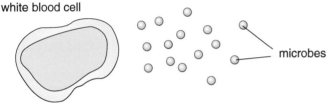

microbes come into contact with white blood cell

microbes

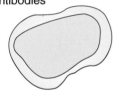 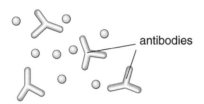

white blood cell releases antibodies

antibodies

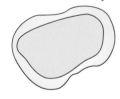 

antibodies react with microbes and destroy them

Protection for life

White blood cells also release chemicals called **antibodies**, which destroy pathogens. For example, when a chicken pox virus enters the body antibodies are made which destroy the chicken pox virus. After the pathogen has been destroyed antibodies remain in the blood to protect your body if the same virus enters in the future.

Once your white blood cells have destroyed a certain type of pathogen, you are unlikely to develop the same disease again. You have become **immune** to the disease.

White blood cells also make **antitoxins** to stop the toxins made by pathogens from poisoning your body.

Questions

a *Name two types of substance that white blood cells produce to protect you from disease.*
b *Choose words from the list to complete the sentence below.*

   white blood cells   red blood cells
   antibodies   platelets   viruses

We are immune to a virus after being infected because _____ produce _____.

Key points

- Microorganisms, such as bacteria and viruses, that cause disease are called pathogens. They produce toxins that make us feel ill.
- White blood cells protect the body against pathogens by ingesting them, producing antibodies to destroy them or producing antitoxins, which counteract the toxins produced by pathogens.
- Semmelweiss was the first person to recognise how to solve the problem of infections spreading in hospitals.

Making you feel better

Once a pathogen gets inside your body you will eventually start to feel ill and show the **symptoms** of disease. The symptoms are the effects the disease has on your body, such as high temperature and headaches.

When you have a disease you can be treated with medicines to make you feel better. For example, people take **painkillers** to ease aches and pains. Painkillers may make you feel better but they do not kill the pathogens that cause the disease.

> **Question**
>
> **a** Give two other symptoms of a common cold, besides a high temperature and a headache.

Killing bacteria

Antibiotics are medicines that help to cure diseases caused by bacteria. You take antibiotics to kill the bacteria that get inside your body. **Penicillin** was the first antibiotic to be discovered.

Antibiotics do not kill viruses. Viruses live and reproduce inside body cells. This makes it difficult to develop medicines to kill viruses without damaging body cells and tissues.

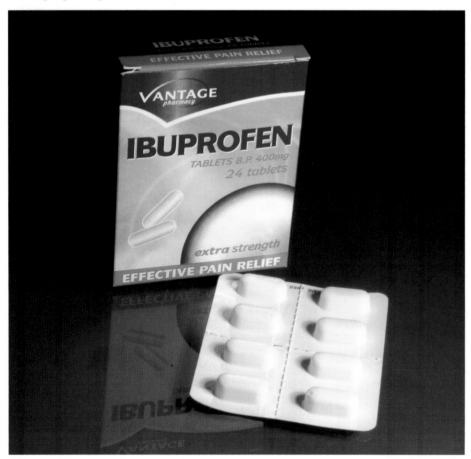

◄ Many medicines are taken to relieve the symptoms of disease.

Superbugs

Some types of bacteria have developed **resistance** to antibiotics – they are no longer killed by antibiotics. MRSA is an example of a disease caused by bacteria that are resistant to antibiotics. When an antibiotic is used, the non-resistant bacteria are killed but a small number of resistant bacteria remain. The resistant bacteria survive and reproduce. This is an example of natural selection. To prevent more and more bacteria becoming resistant, it is important to avoid over-using antibiotics. A doctor will prescribe an antibiotic if it is needed, such as to treat a serious disease.

> **Question**
>
> **b** Measles is a disease caused by a virus. Explain why measles cannot be treated by antibiotics.

Hand hygiene

MRSA is a disease that can affect people who are already ill in hospital. The bacteria that cause MRSA can get onto the skin of doctors and nurses as they treat their patients. The bacteria can then be passed onto other patients. This is why doctors and nurses wash their hands thoroughly as they move from one patient to another.

A study has been carried out to assess the effectiveness of washing hands as a method of preventing the spread of MRSA. Trained staff observed doctors and nurses at timed periods to monitor if they washed their hands and how thoroughly this was done. The graph shows how hand-washing affected the spread of MRSA.

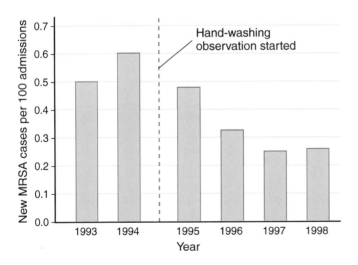

> **Questions**
>
> **c** How many fewer cases of MRSA were there in 1998 compared to 1994?
> **d** Does thorough hand-washing prevent the spread of MRSA? Use the evidence in the bar chart to support your answer.

Key points

- Painkillers relieve the symptoms of diseases but do not kill the pathogens.
- Antibiotics such as penicillin can cure bacterial diseases by killing the bacteria. They do not work on viruses.
- Many bacteria such as flu and MRSA have become resistant to antibiotics due to natural selection, allowing the strongest bacteria to survive.
- Due to an increased understanding of antibiotics and immunity, the way we treat disease has changed.

A quick jab

When you were a young child you were probably **immunised** to protect you from certain harmful diseases, such as whooping cough, measles and polio. Immunisation usually involves injecting or swallowing a **vaccine**. A vaccine contains small amounts of dead or inactive forms of the pathogen that causes a disease. Because the pathogen is weak or inactive, the vaccine does not make you ill but your white blood cells produce the antibodies to destroy the pathogen. This makes you immune to future infection by the pathogen.

Question

a Choose words from the list to complete the sentence.
    antibiotics   antibodies   antitoxins   weakened pathogens
                 white blood cells
    The whooping cough vaccine contains _____, which stimulates the body to make new _____.

Immunisation programmes

Immunisation provides protection against several diseases that used to be very common in children. An example is the use of the MMR vaccine – a combined vaccine that develops immunity to measles, mumps and rubella.

Vaccines such as MMR have saved millions of children from illness and even death. The graph shows the effectiveness of the immunisation programme against measles.

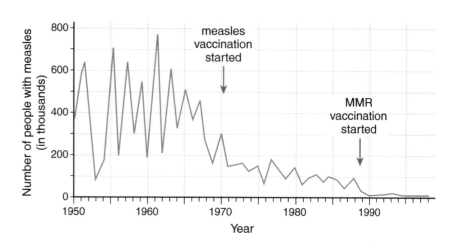

Questions

b What was the maximum number of cases of measles in any one year before a vaccine against the disease was introduced?

c What was the maximum number of cases of measles in any one year after the introduction of the measles vaccine?

Concern about vaccines

Children who are not vaccinated are much more likely to develop serious illnesses.

In the 1970s, many parents became worried after reading newspaper reports about the possible side effects of the whooping cough vaccine. As a result fewer children were vaccinated against whooping cough. Whooping cough is a very serious illness that makes children very ill and can even kill them.

The effect of not having children vaccinated against whooping cough can be seen in this graph.

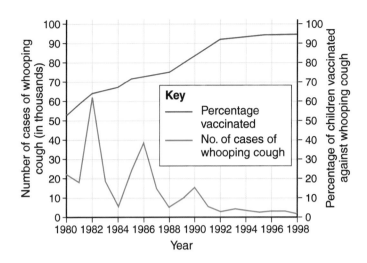

Questions

d In what year did the numbers being vaccinated against whooping cough begin to fall?

e Explain why the number of children catching whooping cough in the 1990s was very low.

Concerns about MMR

Scare stories about the safety of the MMR vaccine appeared in newspapers and on TV. Many parents found it difficult to know who to believe and some parents decided not to have their children vaccinated with MMR. Children who have not been given the MMR vaccine are not protected against measles, mumps and rubella.

The graph shows the percentage of children receiving the MMR vaccination.

Mumps epidemic hits students

Doctors are reporting a big increase in suspected cases of mumps in schools and universities compared to recent years.

More than two-thirds of the students affected missed out on the MMR jab as children.

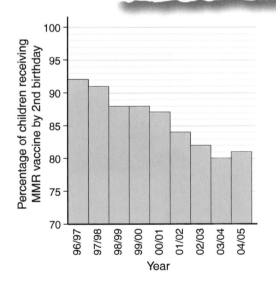

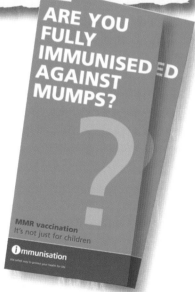

ARE YOU FULLY IMMUNISED AGAINST MUMPS?

MMR vaccination
It's not just for children

immunisation

Questions

From 1996 to 2003 there were many reports in the media about the safety of the MMR vaccine.

f How did these reports affect parents' trust in the safety of the MMR vaccine? Use the graph to explain your answer.

g Why did fewer parents have their children vaccinated with MMR in the late 1990s?

Key points

● People can be immunised using dead or inactive forms of a pathogen in a vaccination. This stimulates the white blood cells to produce antibodies and makes the body immune.

● MMR is an example of a vaccine. Some vaccinations may appear to have side effects and so it is necessary to weigh up the advantages and disadvantages of being vaccinated against a particular disease.

Fighting flu

The poster shown opposite is used to encourage people, especially the elderly, to get vaccinated against flu. Most people recover from flu but elderly people may become seriously ill if they get flu. Unlike other diseases people need to receive a new vaccine against flu each year.

Question

a *Use the information in the poster to explain why people need to be vaccinated against flu every year.*

IF YOU KNEW ABOUT FLU YOU'D GET THE JAB.

The flu is not a severe cold. It's clever. It continually evolves and mutates, so this year's virus may be different from last year's. If you suffer from certain chronic illnesses or you are 65 or over, you are especially at risk. Contact your local GP for this year's free flu jab.

NHS

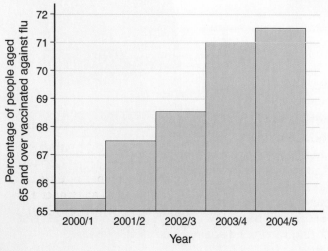

Question

b *The graph shows the percentage of people aged 65 and over who were vaccinated against flu during recent years. What was the increase in the percentage being vaccinated between 2000/1 and 2004/5?*

BIRD FLU KILLS 50TH HUMAN VICTIM

BIRD FLU BIGGER THREAT THAN TERRORISM

A global threat

Flu is caused by a virus. It affects many people every year. Most people recover quickly, but flu can cause serious illness and death, especially in very young children and old people. When an outbreak of flu affects thousands of people in a country it is called a **flu epidemic**.

A **pandemic** outbreak of flu occurs when the virus spreads very rapidly around the world, affecting people in many countries.

Changing viruses

Major outbreaks of flu occur when a new flu virus is produced that is very different from previous strains. Because the new strain is so different, people have no immunity to it. This allows the new strain to cause more serious illness and to spread quickly from person to person.

This graph shows the number of deaths from flu in a country.

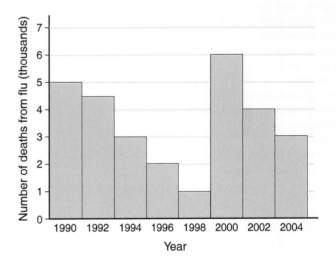

Questions

c How many people died from flu in this country in 1990?
d Calculate the reduction in the number of deaths between 1990 and 1998.
e The flu virus mutated to produce a new strain. In what year did the new strain emerged? Use evidence from the graph to explain your answer.

Bird flu

A number of Asian countries have recently been affected by bird flu. The virus that causes bird flu has infected a small number of people as well as thousands of birds. Scientists are worried that the virus could eventually combine with a human flu virus. This could produce a new virus that is very different from other flu viruses. This could cause a deadly pandemic.

Question

f Explain why a new virus that is different would spread very rapidly.

Key point

- The mutation and resistance of bacteria and viruses makes treatment of illnesses difficult and can lead to epidemics and pandemics, for example bird flu or MRSA.

1 The table is about the receptors we use when we visit a take-away.

Match words from the list with the numbers **1–4** in the table.

   A eye   **B** nose
   C skin   **D** tongue

Structure	Contains receptors which allow us to
1	read the price list
2	feel how hot the coffee cup is
3	taste the burger
4	smell the food cooking

2 The table shows the daily water gain and water loss for a student on a cold day.

The total water gain is equal to the total water loss.

Water gain in cm³		Water loss in cm³	
food	800	sweat	500
drink	1200	breath	400
respiration	400	urine	1400
		faeces	

a How much water was lost in the faeces?
   (1 mark)

b What proportion of water gain was by way of food?   *(1 mark)*

c Which organ produces sweat?   *(1 mark)*

d The student spends a day playing beach volley ball.
   i   What will probably happen to the volume of urine she loses?   *(1 mark)*
   ii Explain the reason for your answer.
       (2 marks)

3 A student accidentally touches a drawing pin. His hand quickly moves away from the pin.

The diagram shows the parts involved in this reflex action.

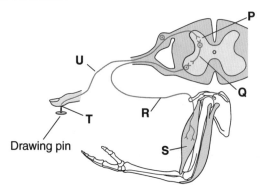

a In this reflex action the receptor is found at
   A Q   **B** R   **C** S   **D** T

b In this reflex action the sensory neurone is found at
   A R   **B** S   **C** T   **D** U

c In this reflex action the relay neurone is found at
   A P   **B** Q   **C** R   **D** U

d Information is NOT carried by electrical impulses at
   A P   **B** Q   **C** R   **D** U

4 This is part of the report of an investigation done by a student.

The purpose of this experiment was to determine the effect of light intensity on sharpness of vision. My hypothesis was that the best amount of light to see in would be a medium-bright light, about 245 lux, because I find it easiest to see when a light isn't too dim, but it's not so bright that I can't work very well.

To measure the responding variable each subject read the letters on an optician's wall chart.

I did the experiment in a windowless room.

I put a chair 6.1 metres away from the wall chart.
I set up a rheostat to dim the lights.
I used a light intensity meter to measure the amount of light reflected off the wall chart in each light level.
I got 20 classmates to volunteer.
I adjusted the light so it reflected from the chart at 780 lux.

A student sat in the chair and read the letters on the wall chart as far as possible.
The experiment was repeated at 759 lux, 243 lux, 81 lux, and 9 lux.
The experiment was then repeated for all remaining students.
My results are shown in the table.

Student	Number of errors made by each student at each light intensity				
	759 lux	243 lux	81 lux	27 lux	9 lux
1	2	1	2	5	5
2	3	3	7	7	8
3	0	0	1	2	7
4	3	4	2	6	7
5	0	1	3	5	12
6	3	2	1	6	10
7	1	2	1	2	3
8	7	7	8	8	15
9	5	4	3	5	8
10	2	0	1	1	3
11	3	3	4	7	8
12	0	1	0	0	4
13	4	4	5	6	13
14	3	2	2	3	2
15	3	4	5	4	13
16	2	1	3	3	4
17	2	4	1	5	10
18	4	5	6	6	12
19	0	1	0	0	3
20	0	1	4	4	5

a   What kind of variable is
   i   reflected light intensity   *(1 mark)*
   ii   number of errors?   *(1 mark)*

b   i   Name one factor that the student
      controlled.   *(1 mark)*
   ii   Name one factor that the student
      did not control.   *(1 mark)*

c   The student used 20 volunteers. Why was
   this better than using 5 volunteers?   *(1 mark)*

d   Work out the average number of errors
   at each light intensity. Write your results
   in a table.   *(2 marks)*

e   Describe one pattern you can see in the
   results.   *(1 mark)*

f   Describe one way of showing these results
   graphically. Say whether you would use a
   bar chart or a line graph, and say what you
   would plot on each axis.   *(3 marks)*

g   Is there any evidence to suggest that the
   student's hypothesis is correct, that the
   best amount of light to see in would be a
   medium-bright light, about 245 lux? Explain
   your answer.   *(2 marks)*

5   The passage contains information about the
   'morning after' pill.

What does the pill do?
The 'morning after pill' stops you from becoming pregnant.
It's not 100 per cent effective, but the failure rate is quite low
– probably about 10 per cent, and rather better than that if you
take it as early as possible.
The pill is believed to work principally by preventing your
ovaries from releasing an egg, and by affecting the womb lining
so that a fertilised egg can't 'embed' itself there.
In the UK and many other Western countries, it is not legally
regarded as an abortion-causing drug, but as a contraceptive.

Who is the pill for?
It's now very widely used by women (especially young women)
who have had unprotected sex. And in particular, it has proved
of value to rape victims, couples who have had a condom break
and women who have been lured into having sex while under
the influence of drink or drugs.

Is it dangerous to use?
You might feel a little bit sick after taking it, but only about 1
woman in every 60 actually throws up. Uncommon side-effects
are headache, tummy ache and breast tenderness.

**If the pill didn't work, and I went on and had a baby, could
the tablet damage it?**
We simply don't know the answer to this question. At present,
no one has shown any increase in abnormalities among babies
who have been exposed to the morning after pill. But past
experience does show that other hormones taken in early
pregnancy have harmed children.

a   Some people regard this pill as an abortion-
   causing drug. Explain why.   *(2 marks)*

b   i   Some people think that this pill should
      only be available on prescription. Suggest
      why they think this.   *(1 mark)*
   ii   Others say it should be freely available
      'over the counter'. What do you think?
      Give reasons for your answer.   *(2 marks)*

c   Scientists are uncertain whether the pill
   might cause abnormalities among unborn
   children. Suggest why.   *(2 marks)*

6   The table is about the effects a poor diet can
   have on the body.

   Match words from the list with each of the numbers
   1–4 in the table.
      A   being overweight
      B   eating too little food
      C   eating too much saturated fat
      D   eating an imbalanced diet

	Effect on body
1	reduced resistance to infection
2	arthritis
3	high levels of blood cholesterol
4	deficiency diseases

7 Adults should eat no more than 6 g of salt a day. You can work out how much salt and fat there is in foods by reading the label. The amount of salt is usually given as the amount of sodium.

Amount of salt = amount of sodium × 2.5

The label on a take-away meal has the following information:

	100 g provides
Fat	8.0 g
of which saturates	6.7 g
polyunsaturates	0.3 g
Sodium	0.6 g

The mass of the whole meal is 300 g.

a Calculate the total amount of
  i saturated fats *(1 mark)*
  ii salt in this meal. *(1 mark)*

b Why is eating a lot of salt bad for your health? *(1 mark)*

c Explain why eating too much saturated fat increases the risk of heart attack. *(3 marks)*

8 The table shows the results of a survey of over 850 adults living the UK in 2001. The survey was carried out by the Food Standards Agency and the Department of Health.

	Percentage overweight by age group				
	19–24	25–34	35–49	50–64	All (19–64)
women	24	28	31	41	32
men	25	42	45	46	42

a Which sex shows the largest percentage of overweight adults? *(1 mark)*

b Which age groups in men and women show the greatest increase in the percentage of people overweight? *(2 marks)*

c Give **two** reasons why these results are reliable. *(2 marks)*

d Being overweight increases the risk of certain diseases. Name **three** diseases linked to excess weight. *(3 marks)*

e Explain why the percentage of overweight adults increases with age. *(3 marks)*

9 The table is about the effects of some substances on the body.

a Match words from the list with each of the numbers **1–4** in the table.
  A alcohol   **B** carbon monoxide
  C nicotine   **D** tobacco

Substance	Effect on body
1	may cause lung cancer
2	may cause damage to liver and brain
3	makes you want to carry on smoking
4	reduces birth weight of babies whose mothers smoke

The graph shows the smoking habits of men and women in 1974 and 2003.

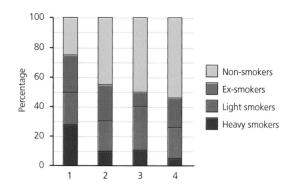

b Match the numbers **A**, **B**, **C** and **D** with the statements **1–4** in the table.
  A 10   **B** 20
  C 50   **D** 60

1	The percentage of women who were ex-smokers in 2003.
2	The percentage of men who smoked in 1974.
3	The percentage of men who were heavy smokers in 2003.
4	The percentage of women who were not smoking in 1974.

10 New drugs must be tested before use. A form of ultrasound is being used by scientists to test the effectiveness of drugs designed to break down potentially life-threatening blood clots. Scientists from King's College of Medicine in London claim the technique provides a more reliable measure of the effectiveness of drugs than was previously available, and could remove the need to test new drugs on animals.

They have used the technique to test the effectiveness of a new drug – GSNQ – which dissolves blood clots. This reduces the risk of strokes. GSNQ was compared with the standard treatment of aspirin and heparin in a group of 24 patients who underwent surgery to clean a major blood vessel in the neck. Patients treated with GSNQ were found to have significantly lower numbers of clots during a three-hour period after the operation.

A member of the research team said: 'Before this technique assessing a drug meant either doing animal tests, or taking blood from people and studying it under the microscope. Neither was a very good measure of what would actually happen when the drug was used in people.'

New drugs will still have to be thoroughly assessed in large-scale clinical trials, but the new technique will help scientists to decide which products should go to a full trial.

a Explain why new drugs have to be tested before they go on sale.   *(1 mark)*

b How did the scientists measure the effectiveness of GSNQ?   *(1 mark)*

c Give **three** advantages of the above method of testing GSNQ over traditional drug-testing methods.   *(3 marks)*

d Explain why GSNQ will still need to be assessed in large-scale clinical trials before it is approved.   *(2 marks)*

11 The table is about the prevention of disease.

Match words from the list with each of the numbers **1–4** in the table.

A MMR vaccine   B MRSA
C painkillers   D white blood cells

	Function
1	strain of bacteria resistant to antibiotics
2	relieves some of the symptoms of disease
3	protects against rubella
4	produces antitoxins

12 The bar chart shows the number of cases of influenza (flu) in a large city in the UK.

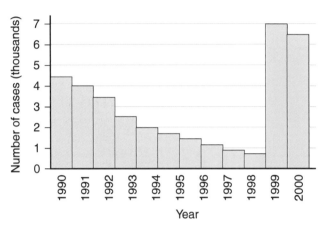

a The decrease in the number of cases of flu between 1990 and 1999 occurred because more people
A became immune to the flu virus
B died from flu
C received antibiotics
D became infected with other diseases.

b The large number of cases in 2000 was likely to have occurred because
A a different type of flu virus was produced by mutation
B more people became immune to the flu virus
C more people were vaccinated against flu
D there were more old people.

c Most people who get flu recover in a few weeks. This is because their white blood cells destroy the virus by producing
A antibodies
B antitoxins
C enzymes
D vaccines.

d When people are vaccinated against flu they receive a vaccine which contains
A antibiotics
B antibodies
C inactive viruses
D resistant bacteria.

Variation and evolution

Living in the Antarctic is impossible for humans without shelter because of the extreme cold and high winds all year round, but penguins live and breed there.

▲ The only way humans can survive in the Antarctic.

Penguins can live in the Antarctic because their bodies have become suited to conditions there. Other animals are adapted for survival in very hot, dry environments.

Adaptations are ways in which organisms have become specialised to survive in a particular habitat. These adaptations result from changes in genes. Competition is another factor that affects survival. Some adaptations enable an organism to compete more successfully. The genes for successful adaptations are passed on to future generations. This is how organisms evolve.

▶ Permanent residents.

Changing genes

Scientists can now change the genetic make-up of plants and animals using genetic engineering. Organisms that have had their genes changed are called genetically modified (or GM) organisms. New varieties of plants and animals have been developed with genes transferred from a completely different species. Bacteria with added genes are used in the manufacture of drugs, such as insulin. Genetically modified crops (GM crops) are grown to increase food production.

Some people think that GM crops can increase the amount of food that can be produced, especially in developing countries in Africa and Asia. However, other people believe that GM food is unsafe and that it is wrong to alter the genetic make-up of organisms.

Preventing inherited disease

In future it may be possible to use genetic engineering to prevent babies being born with an inherited disease. This would involve changing the genes in sperm and fertilised eggs. This could lead to people choosing other characteristics of their children – creating 'designer babies'.

Deciding whether something is right or wrong is an ethical question. Genetic engineering is controversial because it raises many ethical questions.

▶ Genes can be changed in sperm cells and fertilised eggs so that organisms develop with the desired characteristics.

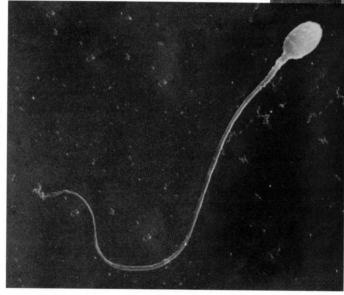

▲ Public opinion remains divided over the issue of GM food.

Think about what you will find out in this section

How are organisms adapted to their surroundings?	How are characteristics inherited?
How can we produce plants and animals with the characteristics we prefer?	What are the advantages of cloning tissues and embryos?
How can the genetic make-up of plants and animals be changed?	What are the arguments for and against changing the genetic make-up of organisms?
Why have some species of plants and animals died out, and new species developed?	Why are there different theories to explain evolution?

Competing for food

These wild dogs are lucky – they will get to finish this meal. But this scene is becoming rarer. Two hundred years ago, wild dogs were common across the whole of southern Africa. As more of the continent becomes used for farming, there are fewer antelopes to feed on. Wild dogs also face competition for food from hyenas and lions. The wild dogs usually lose – their kills are often stolen by the hyenas and lions.

> **Question**
>
> **a** *Suggest why hyenas and lions are often able to steal the kills of wild dogs.*

▲ The wild dogs have killed the antelope – but will they get to eat it?

Competing for territory

▲ Marking his territory.

Lions live in family groups. Each family group has at least one mature male. His job is to stop other lion families competing for the food needed by his family. He marks his territory by spraying urine onto bushes. This acts as a warning to other male lions. If any do stray past the markings there will be a fight – and the winner will take charge of the territory.

Competing for mates

▲ The strongest male mates with the females.

A family group of lions will contain many females but only one or two mature males. These are the males who will mate with the females. Other males are chased away. The only way a male lion outside the group can get to mate is to fight. The winner of the fight becomes the new dominant male of the group and is able to mate with the females.

Even fleas compete

Rotifers and water fleas are tiny aquatic animals. They are sold by pet shops for feeding goldfish.

Two flasks were set up to study competition between these two animals:

- Flask X held only rotifers
- Flask Z held both rotifers and water fleas.

Both animals feed on the same microscopic plants, which were added to both flasks daily.

The graph shows the results of the experiment.

Number of rotifers (y-axis: 0 to 4500)
Day (x-axis: 0 to 18)

▲ How the numbers of rotifers changed over 18 days.

Questions

b Describe what happened to the population of rotifers:
   (i) when the rotifers were living alone
   (ii) when the rotifers were living with water fleas.

c Water fleas do not eat rotifers. Suggest an explanation for the fall in the number of rotifers in flask Z after 10 days.

Competing plants

Plants make their own food from carbon dioxide and water. The energy for this comes from light. They also need some nutrients from the soil.

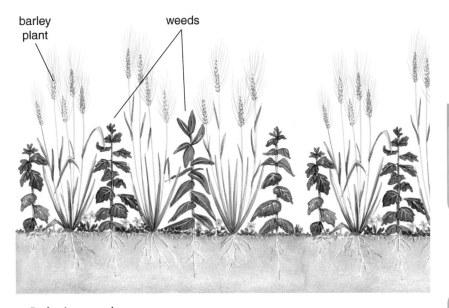

barley plant   weeds

▲ Barley is a cereal crop.

The diagram shows weeds growing in a barley crop. Weeds compete with crop plants for light, water and nutrients from the soil.

Question

d Look at the diagram.
   (i) What advantage do barley plants gain by growing taller than the weeds?
   (ii) What do the barley plants and weeds compete for in the soil?

Key points

- Animals compete with each other for food, territory and mates.
- Plants compete with each other for light, water and nutrients.

Do size and shape matter?

A bigger object stays warmer for longer than a small one. For example, a large cup of tea cools much more slowly than a small cup of tea. It is the same with animals. For animals with the same shape, the larger one will keep its body temperature more easily. Animals lose heat through their body surface so if they can reduce the area of their body surface, they can reduce the rate of heat loss.

A walrus has about the same mass as a horse but it has a completely different shape.

Question

a   *Think about the body shapes of a walrus and a horse. Which parts of a walrus are much smaller compared to the rest of its body than similar parts of a horse?*

▲ Walruses are well adapted to living in the cold Arctic seas.

Huddling

In winter the Antarctic temperature drops to −30 °C and the wind speed can reach 200 km/h. In these conditions penguins huddle together in large groups, making a warm penguin blanket to shield them. They all take a turn on the outside though! The penguins keep warm because the surface area of the whole group is reduced – it is just like having one very large penguin.

▶ Penguins huddle together for protection against the wind.

Insulation

The walrus has a very thick layer of fat called blubber under its skin. Blubber is a very good insulator. In winter this layer is about 10 cm thick – not much heat escapes through that.

Musk oxen live in the Arctic. They are the size of cattle, but have wool, like sheep.

The woolly coat traps air next to the body. Air is a poor conductor of heat. That is why double-glazing windows work – the two panes of glass trap an insulating layer of air.

▲ Musk oxen have long wool that hangs almost to their feet.

Camouflage

Snowy owls live in the Arctic. They feed on small mammals such as mice. In summer, snowy owls are brownish with dark spots and stripes but in winter they are completely white.

Question

b *How is being white in winter an adaptation to life in the Arctic?*

▲ The snowy owl changes its plumage in winter.

▲ Polar bears are well adapted to life in the Arctic.

Question

c *Look at this photograph of a polar bear and explain how the polar bear is adapted for life in the Arctic through its:*
(i) camouflage (ii) insulation (iii) reduced surface area.

Key points

- Animals may be adapted for living in cold places by reducing heat loss by reducing their surface area (for example with smaller ears and shorter limbs), by insulation (for example long hair or a thick layer of fat) or by behaviour (for example huddling to reduce the total surface area of all the animals).
- Animals may also be camouflaged by having white coats in winter.

▲ Cacti can survive in Death Valley.

Death Valley

Death Valley is in California, USA. It is one of the hottest and driest places on Earth.

Cacti

The main problem facing desert plants is dehydration. Heat from the sun evaporates water from their surfaces. A cactus has a swollen stem. The outside of the stem has a waxy covering. Cacti leaves have been reduced to spines. Their roots spread out over a wide area.

Question

a Which feature(s) of the cactus enables it to:
(i) collect as much water as possible when it rains
(ii) store water
(iii) reduce water loss
(iv) prevent animals eating it?

Ears are not just for hearing

The first Europeans who saw elephants assumed that elephants had big ear flaps so that they could hear better. Then some scientists noticed that elephants flapped their ears more at midday when the temperature was at its highest. Further investigation found that the ears had a very good supply of blood. The elephants were flapping their ears to cool down.

We cool ourselves down by sweating, but sweating uses up precious water. The elephant uses its big ears to cool down. Its ears receive warm blood from its body. The warm ears radiate heat to the environment. The cooled blood the returns to the body. This method of cooling conserves water, an important adaptation for a hot climate.

▲ Animals in many other hot parts of the world cool down by having large ears.

Question

b Being big is a good thing in the Antarctic. Explain one disadvantage of being big in a hot climate.

Warning signals

If you see an insect with yellow and black stripes moving towards you, you will probably try to get out of the way. This is because you know that many insects with these coloured stripes sting. Birds know that too – and many bird species will not try to eat them.

Wasp

Hoverfly

▲ Keep-off colours.

Question

c The hoverfly shown in the photograph does not sting, but birds do not try to eat it. Suggest an explanation for this.

Many animals and plants defend themselves by producing poisons. However, they need to advertise this or they will be eaten anyway! Most poisonous animals have brightly coloured markings to warn off predators. The predators themselves have adapted to recognise these colours as a sign of danger.

◀ Native Indians use this frog to make poisoned arrows.

Question

d The Io moth uses a different method to warn off predators. Suggest why many birds would not attack this moth.

◀ Io moth.

▲ Sharp defences stop cows from eating thistles.

Cows know to eat the grass, but ignore the thistles because they will be stung.

Question

e Suggest why the thistles do not get eaten.

Key points

- Plants survive in hot places by having a thick, waterproof covering, reducing the area of their leaves, storing water and having long roots.
- Animals survive in hot places by increasing their surface to increase heat loss.
- Animals and plants may have thorns, poisons or warning colours to deter predators.

A quick look at this photograph of a lawn will tell you that the plants with flowers are not distributed evenly across it. Because of this it is not easy to find out how many of each type of plant there are. It might be possible to count all the daisies on a small lawn, but that would be impossible for the school playing fields.

Instead of trying to count all the plants we can use **sampling**. We count the number in a small area, then use this number to estimate the total.

Question

a *In making this estimate, what are you assuming about the area you have used and the rest of the lawn?*

Quadrats

The most common method of sampling animals and plants is the quadrat. This is a square frame, usually either $50\,cm^2$ or $1\,m^2$. Quadrats are often subdivided into 10-cm squares.

Five students each threw a quadrat onto the lawn and counted the number of daisy plants in the quadrat. The diagram (below right) summarises their results.

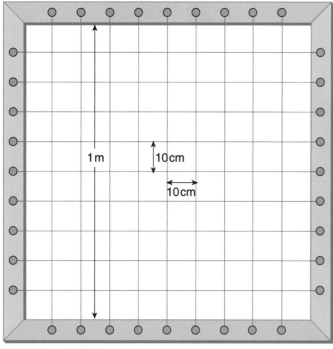

▲ 1-m quadrat divided into 10-cm squares.

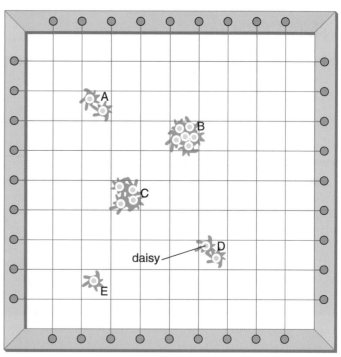

▲ Counting the daisies.

Question

b (i) *Suppose the students had only thrown quadrat A. What would the estimated total population of daisies on the lawn have been? What would it be for quadrat B?*
(ii) *Find the average number of daisies in the five quadrats then estimate the total number of daisies on the lawn. Compare this with your answer to part (i). What does this tell you about sampling?*

When conditions change

The distribution of organisms in a habitat often depends on the different conditions there. Conditions which may vary in a habitat include light, shade, water, pH, **ions** and temperature. Any or all of these may affect the distribution of the organisms.

The diagram shows a quadrat placed next to a hedge.

Count the number of plants in each of the rows 1–10. Do not count any plant more than once. Record your results in a suitable table. Now draw a bar chart of your results. Do not forget to label the axes.

Question

c (i) Use your bar chart to describe the relationship between the number of plants and the distance away from the hedge.
(ii) Suggest one hypothesis to explain this relationship.
(iii) Outline how you would plan an investigation to test this hypothesis. Remember to include:
   – the independent variable
   – the dependent variable
   – how you intend to measure the dependent variable
   – the control variables.

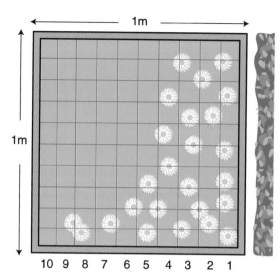

▲ Distribution of plants near a hedge.

Using a transect

If conditions change over several metres, scientists use **transects** to place quadrats. A transect is a measuring tape running across the area we want to investigate.

The diagram shows the students' results for two species of snail: the small periwinkle and the rough periwinkle. The length of each band shows where the periwinkles were found. The breadth of each band is proportional to the numbers of each found.

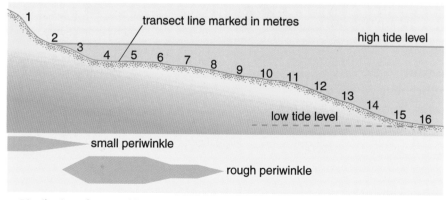

▲ Distribution of periwinkles on a sea shore.

Question

d (i) Describe the distribution of the two species of periwinkle.
(ii) Suggest one reason for the different distributions of the two periwinkles.

Key points

● We can use transects and quadrats to sample the distribution of organisms in a habitat.
● The larger the number of quadrats, the more reliable the results.
● The distribution or organisms in a habitat is affected by many factors, including light, shade, temperature, water, ions and pH.

There's no one like you!

There may be many people in your school but you can easily tell them apart. Even though there are millions of people in the world, no two are exactly the same. Height, eye colour and earlobe shape are just three **characteristics** that vary from person to person.

Parents and their children often look similar because many characteristics are **inherited** – they are passed from parents to their offspring. Young animals resemble their parents because these characteristics are passed on to them.

A group of students recorded some of the different characteristics in their class. The results are shown in the table.

Sex	Type of ear lobe	
	Attached	Free
Female	12	3
Male	10	0

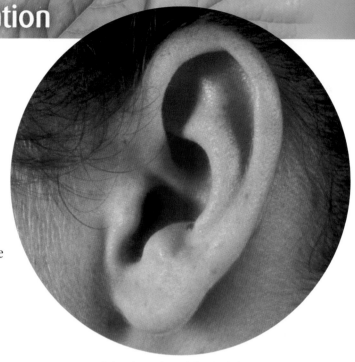
▲ Earlobe shape is an inherited characteristic.

Question

a All of the students with free earlobes were girls. The students concluded that boys do not have free earlobes. Explain why this conclusion is not justified.

Passing on genes

Inherited information is carried by **genes**. Different genes control different characteristics. All your genes are carried on the **chromosomes** found in the nucleus of your body's cells.

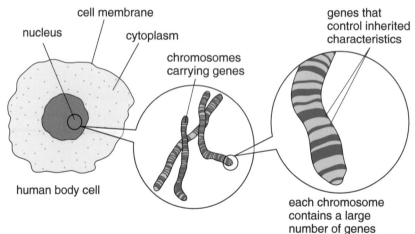

The start of life

Your life started when a **sperm** cell from your father fertilised an **egg** cell from your mother. Egg and sperm cells carry genes from the parents. You inherited half of your genes from your father and half from your mother.

Sperm and egg cells are specialised cells called **gametes**. The joining (fusion) of male and female gametes is called **sexual reproduction**. Plants also produce gametes in their flowers.

Sexual reproduction produces offspring with a mixture of genetic information from both parents.

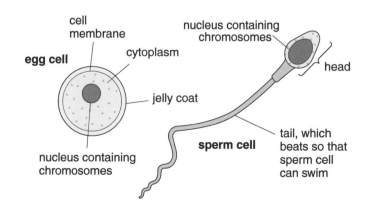

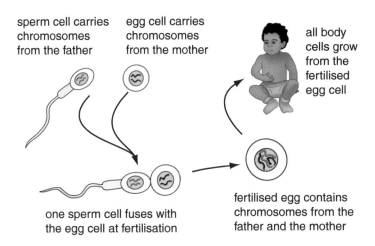

sperm cell carries chromosomes from the father

egg cell carries chromosomes from the mother

all body cells grow from the fertilised egg cell

one sperm cell fuses with the egg cell at fertilisation

fertilised egg contains chromosomes from the father and the mother

Question

b Explain why children show a mixture of characteristics from both parents.

Reproducing from one parent

Some plants and animals can reproduce by **asexual reproduction**. In this type of reproduction there is no fusion of cells and only a single parent is needed. The diagram shows new strawberry plants forming from one parent plant. All the individuals produced by asexual reproduction have exactly the same genes as the parent. Organisms with identical genes are known as **clones**.

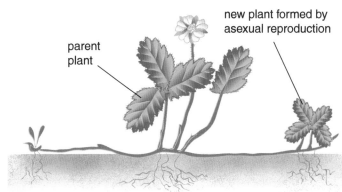

parent plant

new plant formed by asexual reproduction

The photograph shows bean aphids feeding. In the autumn male and female aphids mate and produce fertilised eggs. In spring these eggs hatch. The young aphids are all female. These females then produce offspring without needing any males.

Questions

c (i) Name the type of reproduction that produces aphid eggs in autumn.
(ii) Explain why the females that develop from these eggs are not identical.

d Name the type of reproduction that produces offspring from females without needing any males.

Key points

- Information is carried from parents to offspring in genes.
- Genes are carried on chromosomes found in the nucleus of cells.
- Sexual reproduction produces a mixture of genetic information from two parents.
- There is no mixing of genetic information in asexual reproduction.

Plants from cuttings

Young plants can be grown from older plants by taking cuttings. Plants grown from cuttings taken from the same plant have identical genes. Their genes are identical to the parent plant and to each other. Plants and animals that have the same genetic make-up are called clones.

A new plant can be grown quickly and cheaply from each cutting, without waiting for sexual reproduction to take place. A disadvantage of producing many identical plants is that a disease could damage or kill all of them very quickly.

Questions

a Why are plants grown from cuttings identical?
b Why do plants grown from seeds show a mixture of characteristics from both parents?

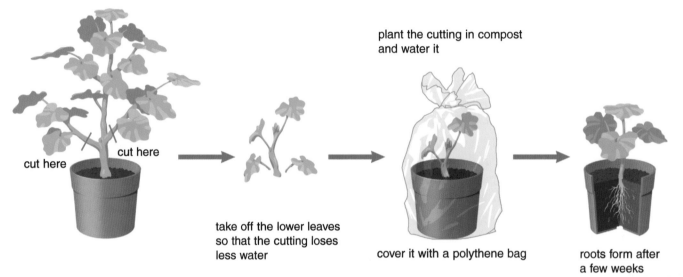

cut here

cut here

cut here

take off the lower leaves so that the cutting loses less water

plant the cutting in compost and water it

cover it with a polythene bag

roots form after a few weeks

Test-tube plants

Plant **tissue culture** is an example of modern cloning. This involves using tiny pieces of plant tissue to grow whole plants. Even large trees can be grown from just a small piece of tissue. Roots, stems and leaves grow from the cells in the piece of tissue. Using tissue culture, plant breeders can grow large numbers of identical plants from just a small piece of tissue. Scientists use tissue culture to reproduce very rare species, especially when a new plant cannot be grown from cuttings.

Question

c Give one advantage of using tissue culture rather than taking cuttings.

Speeding up breeding

Farmers use only animals with the most useful characteristics for breeding. For example, a dairy farmer wants to breed cows that produce large amounts of milk.

By using **embryo transplants** breeders can produce a large number of genetically identical calves from a single fertilised egg. This involves taking eggs and sperm from the best cows and bulls. The diagram shows how identical embryos produced from these gametes are then transferred to host mothers.

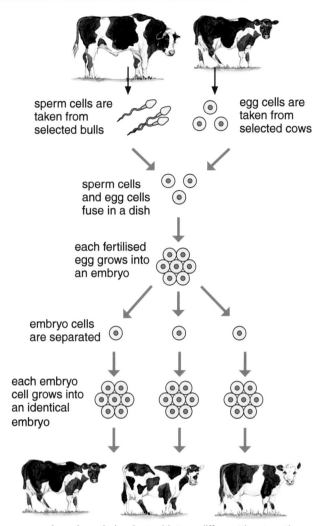

sperm cells are taken from selected bulls

egg cells are taken from selected cows

sperm cells and egg cells fuse in a dish

each fertilised egg grows into an embryo

embryo cells are separated

each embryo cell grows into an identical embryo

each embryo is implanted into a different host mother

Questions

d Where does gamete fusion take place in the process shown in the diagram?

e Why do transplanted embryos carry identical genes?

The world's first copy-cat

The world's first cloned cat was born in 2002. The American company that carried out the cloning called the cat CC, which stands for CopyCat. A woman paid the company $50 000 to clone her dead pet cat. Newspapers reported this story but did not give the woman's name because of fears that she would be targeted by people who were against using animals for scientific research. Why would people be against cloning?

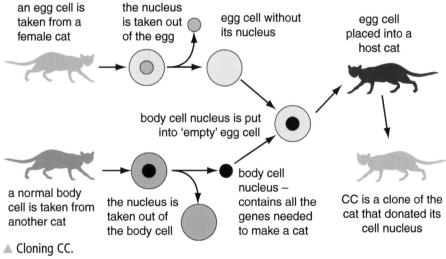

an egg cell is taken from a female cat

the nucleus is taken out of the egg

egg cell without its nucleus

egg cell placed into a host cat

body cell nucleus is put into 'empty' egg cell

a normal body cell is taken from another cat

the nucleus is taken out of the body cell

body cell nucleus – contains all the genes needed to make a cat

CC is a clone of the cat that donated its cell nucleus

▲ Cloning CC.

Question

f (i) Explain why CC has the same genes as the cat whose body cells were used.

(ii) Why didn't CC inherit any genes from the cat that gave birth to her?

Key points

- New plants can be produced quickly and cheaply from cuttings.
- Identical organisms can be produced using modern cloning techniques, including tissue culture, embryo transplants and adult cell cloning.
- Producing plants and animals using cloning techniques raises ethical issues.
- People need to be able to make informed judgements about the ethical issues concerning cloning.

Gene transfer

Scientists have developed ways to transfer genes from one organism to another. This process is called genetic engineering or **genetic modification** (GM for short). Using genetic modification the genetic make-up of an organism can be changed to produce new varieties very quickly.

Making medicines

Bacteria that have had their genes modified can be used to make medicines. For example, bacteria are used to make human insulin. Insulin is a hormone produced by your body to control the amount of glucose in your blood. People with a disorder called **diabetes** produce very little insulin. They need regular injections of insulin to control the amount of glucose in their blood.

Cutting and sticking genes

Insulin used by diabetics used to be obtained from cattle or pigs. But in 1982, scientists began to use genetically modified bacteria to make human insulin. You have a gene that tells your body how to make insulin. Scientists can 'cut' this gene out of the chromosome that carries it and insert it into bacteria.

stage 1

The gene that makes insulin is cut out from the DNA.

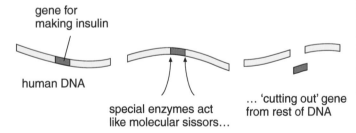

gene for making insulin

human DNA

special enzymes act like molecular sissors...

... 'cutting out' gene from rest of DNA

stage 2

The human gene is inserted into bacterial DNA.

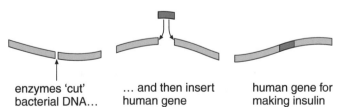

enzymes 'cut' bacterial DNA...

... and then insert human gene

human gene for making insulin

Question

a Suggest an advantage of using human insulin made by bacteria instead of insulin taken from pigs.

Bacterial factories

The bacteria containing the insulin gene are grown inside large vessels. The conditions are controlled so that bacteria multiply very quickly. Each bacterial cell produces a very small amount of insulin, but because rapid growth produces millions of bacterial cells, a large amount of insulin is produced.

cell division

Genetically engineered bacterial cell divides many times, providing numerous identical cells called **clones**

Human insulin is extracted from millions of bacterial cells grown inside bioreactors

Question

b It is important to control the pH inside the vessel that the bacteria grow in. Name another condition that must be controlled to provide the best conditions for growth.

GM crops

Genes can also be transferred from bacteria and added to crop plants. Crops that have had their genes changed in this way are called genetically modified crops (**GM crops**). For example, some bacteria produce a substance that is poisonous to insects. The gene that makes this poison can be cut out of bacteria and added to crop plants. The modified crop plants will now produce the poison. This means that farmers will not need to spray their crops with pesticide to kill insects.

Question

c Explain why farmers using this GM crop will no longer need to use pesticides.

Key points

- In genetic engineering, genes can be 'cut out' and inserted into other organisms.
- Plants and animals will develop with specific characteristics when genes are inserted at an early stage in their development.
- Modifying the genetic make-up of plants and animals raises ethical issues.
- People need to be able to make informed judgements about the ethical issues concerning genetic engineering.

Pet cloning

The first cloned cat was born in 2002. Some people are paying to have the DNA from their dogs and cats frozen in case they want to replace them in the future. Genetic Savings and Clone Inc. is an American company that describes itself as 'The world's leader in the cloning of exceptional pets'. The information on their website shows one of the clients that this company uses to advertise its services.

Cloning companies have 'banked' tissues from the pets of hundreds of customers who want to have their pet cloned when the technology has been fully developed. The pet owners hope that one day the tissues will be used to replicate their pet animal. The director of the company said, "We have banked lots of cells of dogs, cats, horses and cattle. There are many people out there interested in cloning their pets. "The article opposite is the story from one customer who wants her pet cat cloned. Companies use similar articles as part of their website advertising.

Pet cloning – is it a good thing?

A report on pet cloning has recently been published by an animal rights group that has been monitoring the treatment of animals. Animal rights campaigners are against the use of animals for scientific research.

When people have to decide whether something is right or wrong, they are deciding about an ethical issue. Using cloning to produce pets, and to produce human embryos, are examples of ethical issues.

Petsave biotechnology – pet cloning services

My Cuddles lives on...

"Cuddles was my lovely cat. Sadly, cancer got the better of him and I still miss him every day. I know that a little piece of him still lives on in the tissue sample taken from him, and could one day become a new living cat just like Cuddles. I am fascinated that science has made this possible."

Perfect copies

Animal cloning studies have found that only very few cloned animals survive until birth. For example, out of 87 transplanted embryos, CC (see page 55) was the only one to survive to birth. The animals used as host mothers have to undergo surgery for implantation of the embryo. As yet, there has been no long-term study to find out if cloned pets will live normal, healthy lives.

Many pet owners believe that cloning will produce a perfect copy of the animal they love, but even though cloned animals are genetically identical to the donor animal, they are unlikely to be identical in appearance. This is because the appearance of a pet depends on the genes it inherits and the environment in which the animal grows up.

> It would be marvellous to have a cat just like my Sammy. He was a real comfort for me.

Pet owner, whose cat, Sammy, has just died

> Cloning cats has been very difficult. It has taken us years to develop the technology. We can now use it to study and treat diseases in animals and in humans.

Animal rights campaigner

Research scientist

> It is wrong to treat animals in this way. Many cats had to go through painful surgery just to show that a cat can be cloned. Most of the embryos that were implanted died and caused miscarriages.

> People get very attached to their pets. We are offering a service to give pet owners the chance to have the pets they want.

Director of pet cloning company

A vet who treats cats and dogs

> People who have just lost a pet shouldn't be exploited like this. This is wrong for both the animals and the pet owners.

Question

c *Do you agree with pet cloning? Use the information above to give as many reasons as you can to support your view.*

Key points

- The applications of science in medicine and food production can raise social and ethical issues.
- People need to be able to make informed judgements about social and ethical issues concerning cloning and genetic engineering.

Changing species

The huge variety of species that now live on the Earth have not always existed. Species change: new species are formed and others become extinct. There is evidence that all species have evolved from simple organisms that lived on Earth more than 3 billion years ago.

Evidence from the past

Evidence that species have changed can be seen in fossils. Fossils are the remains of plants and animals that lived long ago. They can be found in rocks, tar pits and ice. Fossils form in several different ways:

- hard parts of animals and plants do not decay easily and become trapped in rock

- animals and plants die in cold, dry or oxygen-poor conditions where they don't decay easily

- animal footprints are preserved in mud that becomes rock over thousands of years.

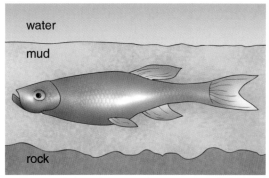

A fish dies and becomes covered by mud.

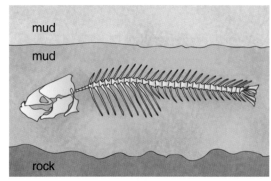

The soft tissues of the fish decay. The only part that remains is the skeleton.

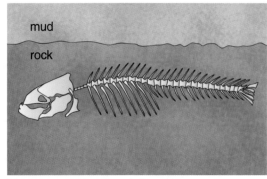

Mud surrounding the skeleton turns into rock.

Question

a Explain why:
  (i) whole bodies of mammoths have been found in frozen soils
  (ii) fish fossils don't show muscles and other soft tissues.

A record of evolution

Fossils provide evidence of how much organisms have changed over time. The position of a fossil helps scientists to identify its age. The deepest layers of rock are likely to contain the oldest fossils. The fossils in layers of rock provide a record that shows plants and animals have changed over a very long period of time.

Fossils are only rarely formed. This is why there are gaps in the fossil record and scientists can only suggest how one kind of organism evolved from another. Not enough fossils have been found to show exactly what happened to each kind of organism that lived in the past. Fossils of simple organisms, such as organisms made from just one cell, are extremely rare. This is one reason why scientists cannot be certain about how life began on Earth.

Extinct species

The fossil record shows that some species lived in the past but are no longer living – they have become **extinct**. Species become extinct because:

- their environment changes
- new predators or diseases kill them
- they cannot compete with other species.

Kakapos are flightless birds found in New Zealand. They are found in forests, where they feed on berries and leaves. Kakapos used to be common but they are now very rare and may even become extinct unless they are protected. Possible reasons for the decline in the numbers of kakapos are:

- farmers have cleared large areas of their forest habitats to provide grazing land for sheep
- rats escaped into the forests and have bred to produce large numbers – rats feed on birds' eggs
- farmers have released deer into the forests, where they eat berries and leaves.

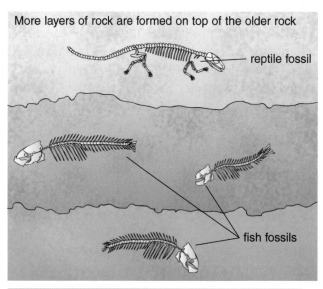

More layers of rock are formed on top of the older rock

reptile fossil

fish fossils

Question

b Use the fossil record above to answer the following questions.
(i) Which group of animals evolved first?
(ii) What evidence supports your answer?

Key points

- Simple forms of life from billions of years ago evolved into the species of all living things.
- Fossils provide evidence of how different organisms have changed.
- Some species have become extinct.
- Extinction may be caused by environmental change, predators, disease or competition.
- Scientists cannot be certain about how life began on Earth.

Questions

c How did the Kakapo's environment change?
d What species competed with the Kakapo for food?

Changing ideas

People used to think that all the species alive today had always existed. As scientists learned more about plants and animals they concluded that species are changing. New species have been formed and others have disappeared.

One of the earliest theories of how these changes take place was proposed by a scientist called Lamarck. According to Lamarck, an organism acquires new characteristics during its lifetime and can pass these on to its offspring. The species changes as each generation passes on useful features. The diagram shows how Lamarck explained how wading bird species developed long legs.

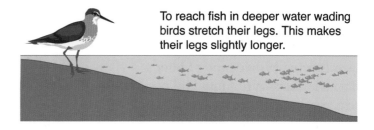
To reach fish in deeper water wading birds stretch their legs. This makes their legs slightly longer.

Having slightly longer legs is passed on to the next generation. Birds in this generation also stretch their legs.

Over many generations, the wading birds' legs become much longer.

Question

a What are the benefits of wading birds having long legs?

Darwin and natural selection

Our present ideas about the way species evolved are based on the **theory of natural selection**, which was first put forward by Charles Darwin. Darwin's theory of evolution states that natural selection brings about changes in species because:

- changes in genes (mutations) produce new forms of genes
- differences in the genes of individuals produce different characteristics
- individuals with characteristics most suited to the environment are more likely to survive and breed
- the genes that have enabled an individual to survive are passed on to the next generation.

Natural selection can only occur when there are differences in individuals of the same species. These differences occur because mutation produces new forms of genes. This is why the change in a species may become more rapid as new forms of genes are produced.

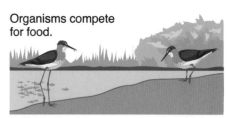
Organisms compete for food.

Individuals of the same species may have different characteristics, such as slightly longer legs.

Individuals struggle to survive. Some die because of lack of food or may be eaten by predators.

Individuals with useful characteristics are more likely to survive, and pass on their characteristics to the next generation.

▲ Darwin's theory of natural selection.

Question

b Darwin's theory of how species change is very different from Lamarck's theory. Explain how wading birds have developed long legs by the process of natural selection.

Different theories

Species have evolved over billions of years. There is not enough evidence to prove beyond doubt why species changed over this very long time period. This why scientists cannot be certain about how life began on Earth and why there are conflicting theories of evolution.

Surviving and breeding

The effects of natural selection can be seen in a species called the peppered moth. There are two varieties of this species – a light variety and a dark variety. The photograph shows the two varieties resting on a tree trunk in a city.

Question

c *Which variety of moth is more likely to be eaten by insect-eating birds? Explain your answer.*

In an investigation:

- large numbers of light and dark varieties of moth were caught in a trap
- the moths were marked with a spot of paint on the underside of their body and then released
- the moths were released into a woodland near a city
- after a few days the moths were trapped again and the number of marked moths was counted.

The results are shown in the bar graph.

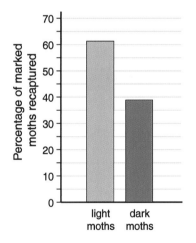

Questions

d *Suggest why the moths were marked with paint on the underside of their bodies.*
e *What percentage of light moths were caught in the woodland?*
f *Use Darwin's theory of natural selection to explain why fewer dark moths were recaptured.*

Key points

- Scientists cannot be certain about how life began.
- There are differences in Darwin's theory of evolution and conflicting theories.
- There may be a more rapid change in a species when genes mutate.
- An investigation determines if a relationship exists between two variables.
- Scientists identify patterns and relationships to make suitable conclusions.

Reacting to new ideas

When Charles Darwin presented his theory of evolution over 100 years ago, people reacted with anger because:

- the theory of natural selection undermined the idea that God made all the animals and plants that live on the Earth as they appeared at the time

- the idea that humans were related to apes such as chimpanzees caused shock and anger.

There is a lot of evidence to support the theory of evolution, including fossil evidence and the similarities and differences between species. Even so some people do not accept that evolution has occurred.

Similarities and differences

Studying the similarities and differences between species provides scientists with more evidence of their evolutionary relationships. The diagrams show the skeletons of an ape, a monkey and a human. The features of the skeletons provide evidence to show how closely related these animals are.

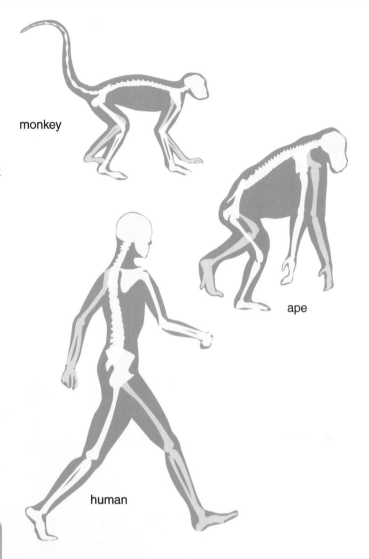

monkey

ape

human

Questions

a Describe two features of monkey skeletons that are not found in the ape skeleton.
b Which animal is more closely related to humans? Give one feature from the diagram to support your answer.

Tracing human evolution

The theory of evolution does not suggest that chimpanzees turned into humans. Chimpanzees are themselves the result of evolution. Evidence suggests that an ancestral ape species gave rise to both chimpanzees and humans. Your family tree shows that you are more closely related to your parents than to your aunts and uncles. Imagine that you could trace this tree back for hundreds of thousands of years. It would show that you shared a very distant ancestor with other apes. The diagram opposite shows a timeline showing the evolution of humans and some other animals.

▲ Recent evidence, such as apes using tools, shows that they are very intelligent, and very closely related to humans.

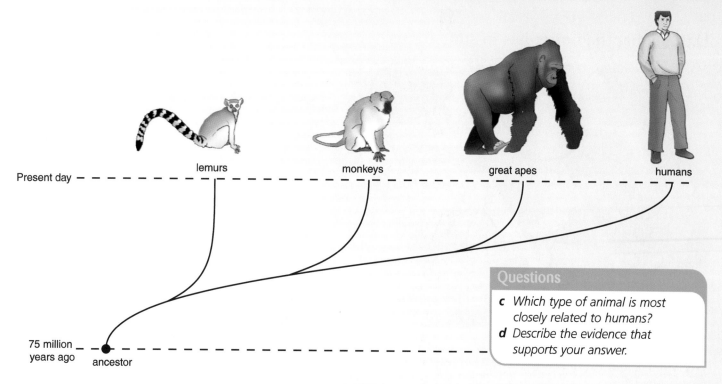

Present day -
lemurs   monkeys   great apes   humans

75 million
years ago - - ● -
        ancestor

Human fossils

Fossil remains provide evidence of how humans evolved. One very important fossil discovery was the bones of a human-like creature that lived 3.2 million years ago. The scientists who found the fossil called it 'Lucy'. The shape and size of the bones provide evidence of what Lucy would have looked like. The structure of the skeleton suggests that Lucy's species is an extinct human ancestor.

There are many bones missing from the skeleton of Lucy. This shows why scientists cannot be certain about how species may have evolved.

Key points

- Darwin's theory of evolution was only gradually accepted.
- Scientists cannot be certain about how life began on Earth.
- The similarities and differences between species may provide evidence of their evolutionary and ecological relationships.

Taking care of the planet

Is this what the future has in store for us?

▲ The 22nd century school run.

endangered species

snowman

▲ Winters in the 22nd century.

The impact that humans are having on our planet is often in the news. Already we are beginning to see climate changes that may bring disastrous changes to many parts of the world. The world is not ours to use as we wish. We hold it in trust for future generations, so we must look after the Earth and its resources.

The photos show some of the concerns that people have about the environment.

▶ Many developing nations are increasing the size of their industries, causing air pollution. Air pollution in China covers an area larger than the UK.

▼ Forests in many parts of the world are being cleared to provide grazing land for cattle. This destroys places where wild animals and plants live. Cutting down forests also impacts the Earth's climate.

▲ Climate change is bringing about more flooding. In recent years scenes like this have become much more common in many parts of the world.

▼ Animals and plants have adapted to their habitats over millions of years. Climate change will cause some habitats to disappear – and with them the animals and plants that live there. Your grandchildren may only be able to see animals like polar bears in a zoo.

Scientists have been measuring environmental change for many years. They use the information that they gather to make predictions about the world in the future.

This is what many scientists think will happen if there is 'business as usual' – if we do not change the way we treat our planet. By 2050:

- global warming will accelerate
- sea levels will rise
- forests will disappear
- tropical diseases will become much more common in Europe and North America
- African and Asian countries may go to war over the rights to precious water.

We are already beginning to see local effects – farmers in Kenya and Ethiopia are fighting over grazing land and water. But climate change is a global problem because air pollution does not recognise boundaries.

Think about what you will find out in this section

How is the growing human population affecting the surface of our planet?	How do scientists investigate the effect of human activities on the atmosphere?
How are our activities affecting the atmosphere and climate?	How can science help us to plan sustainable development?
Why are humans making it difficult for animals and plants to survive?	

4.1 Human population growth

The world's population

As the new millennium began in 2000, the world's population was around 5.5 billion.

Question

a *Look at the graph showing estimates of world population growth. Use it to estimate the population in 3000 AD if there is 'business as usual'.*

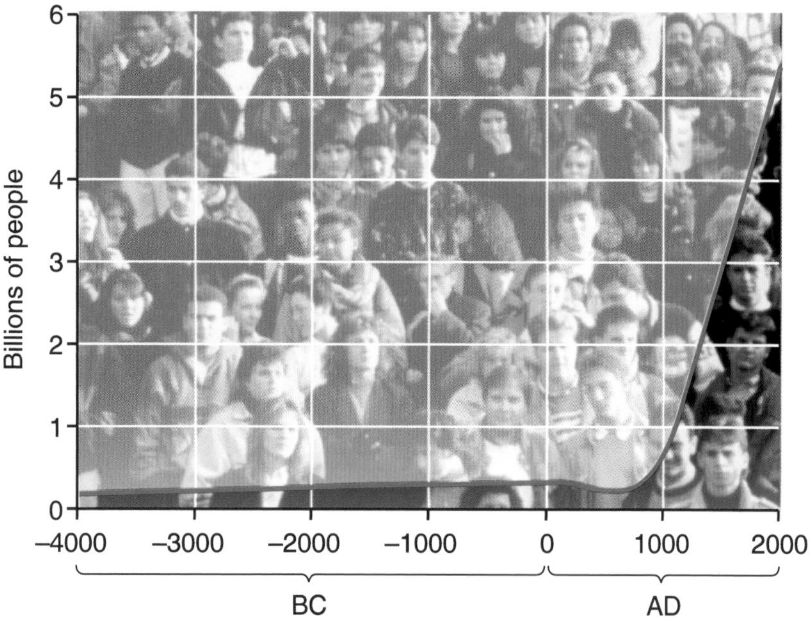

▶ Estimates of world population from 4000 BC to 2000 AD.

Six thousand years ago, the population of the world was about 200 million. People lived in small groups and most of the world was unaffected by human activities.

Running out of raw materials

As world population increases, natural resources are rapidly being used up. Some of these resources, such as fossil fuels, are **non-renewable**. They cannot be replaced, so eventually supplies will run out. As we use raw materials to fuel vehicles and power industries, a great deal of waste is produced. Much of this waste pollutes the environment.

Less space for wildlife

The photo on the right shows a familiar sight in many parts of the world: land being swallowed up for new houses. The growing population means that more and more homes are needed every year. Before building began, this land provided habitats for animals and plants. The building materials came mainly from quarries. Every new quarry destroys habitats.

The inhabitants of each new house will produce waste. This has to be disposed of. Most of the rubbish is dumped in landfill sites. Still more habitats are lost to animals and plants. Poisonous chemicals often drain from landfill sites, killing species that live there.

Question

b *Give three ways in which human activities reduce the amount of land available for other animals and plants.*

▲ New housing swallows up the land.

▼ Shanty towns grow up around cities in many developing countries. These usually have no sewage systems so they are very polluting.

Toxic chemicals

An increasing population means that the world needs more food. Much of the land in the UK is used for agriculture. But using land for agriculture reduces the number of habitats for wild animals and plants.

Farmers use **pesticides** to kill insects that damage their crops. They also spread fertilisers on their crops. If pesticides and fertilisers are washed into streams, they can kill the animals living there.

Butterflies – an endangered species?

Farmers use chemicals called **herbicides** to kill weeds that affect their crops. If these chemicals are washed into streams, ponds or onto meadows, they kill the wild plants living there too. Many of these plants provide food for wild animals such as butterflies.

If the plants at the bottom of a food chain are killed, the **consumers** further up can die out too. The numbers of most kinds of butterflies in the UK have declined dramatically over the last 50 years as farmers have used more herbicides. Herbicides kill flowering plants but not grass crops. Butterflies feed on nectar from flowers – so they have lost much of their food supply.

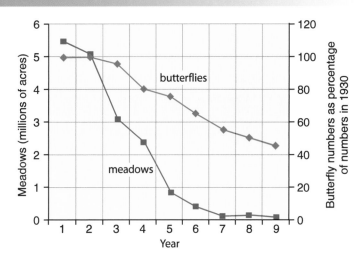

▲ Meadows with lots of wild flowers have nearly disappeared.

Questions

c Is there a direct relationship between the area of meadows and the number of butterflies? Explain your answer.

d Name the type of chemical (i) used to kill weeds (ii) used to kill insect pests.

e Insecticides are only poisonous for insects. But use of insecticides also reduces bird populations. Use your knowledge of food chains to explain this link.

Key points

- Human population growth means that raw materials are being used up quickly and more waste is being produced, causing more pollution.
- Humans are destroying animal and plant habitats by building, quarrying, farming and dumping waste.
- Some species are finding it difficult to survive.

Energy guzzlers

We need energy to power computers, DVD players, washing machines and motor vehicles – all the things that people enjoy using every day, but producing this energy can have serious effects on our local environment.

In 2003, scientists published a report on the link between car fumes and asthma.

They chose 3535 children, aged 9 or over, who did not have asthma. Some of the children lived in cities and the others lived in the countryside. They recorded how much sport these children played.

They found that children who spent a lot of their spare time playing sports in cities were over four times more likely to develop asthma than sporty children who lived in the countryside.

Some forms of asthma are linked to air pollution in cities. Chemicals from car exhaust fumes make up most of this pollution. The fumes contain both gases and tiny particles, both of which may have harmful effects on people. Some of the gases cause acid rain, others contribute to global warming.

Questions

a Why did the scientists study such a large number of children?

b Why do you think that city children who played more sport were more likely to develop asthma than less sporty children who also lived in cities?

Acid rain

All fossil fuels contain molecules called **hydrocarbons**. These are oxidised when the fuel burns, releasing energy. Carbon dioxide is produced as a waste product.

fuel + oxygen → carbon dioxide + water

Petrol contains nitrogen and sulfur compounds. This means that nitrogen oxides and sulfur dioxide are also formed when petrol is burned. All these waste gases pass into the atmosphere. These gases affect cities directly and the waste gases from cities all over the world are changing the Earth's atmosphere.

▲ This boy knows when there are a lot of car fumes in the air – it brings on an asthma attack.

Carbon dioxide dissolves in rainwater and makes it slightly more acidic. Normal rainwater has a pH of between 5.5 and 6.0. Sulfur dioxide from burning fuels dissolves in rain, making it even more acidic. Rain with a pH of 5.2 or less is known as acid rain. Remember that pH 4 is 10 times more acid than pH 5. Acid rain has very damaging effects on the environment. Acid damages living tissues. It can also corrode some types of stone.

Power stations that burn fossil fuels such as coal also release huge amounts of carbon dioxide into the atmosphere and contribute to acid rain.

Question

c Name three gases that may be produced when fossil fuels are burned.

How does acid rain affect the environment?

The photograph shows trees in the mountains of Norway that have been damaged by acid rain. One of the first signs of acid rain damage is that the young leaves near the top of the tree die – this is called 'crown-loss'. Crown-loss is not the only damage caused by acid rain. The acid affects the covering of the older leaves. This makes it easier for disease organisms to attack the tree.

Lichens are living organisms that live on the bark of trees and on stones. Scientists have found that sulfur dioxide kills lichens. Some lichen species are more sensitive than others. The table shows the number of species of lichen on the bark of trees at different sulfur dioxide concentrations.

Question

d Look at the data for lichen and sulfur dioxide.
(i) What is the relationship between the concentration of sulfur dioxide in the air and the number of species of lichen?
(ii) Could you use the number of lichen species to give an accurate prediction of the sulfur dioxide concentration in the air? Explain your answer?

	Concentration of sulfur dioxide in air (micrograms per m³ of air)					
	5	30	35	50	70	150
Number of species of lichen	14	10	8	7	5	2

When acid rain runs into streams and lakes, it makes the water more acidic. Fewer animals and plants can survive. Remember that vinegar has a pH of 3 and lemon juice a pH of 2. The bar chart shows how the pH of water affects which animals live in it. All the animals listed, except snails, are fish.

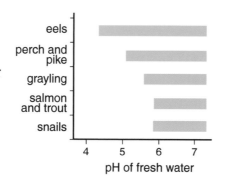

Question

e (i) Which fish copes with acid water the best?
(ii) How reliable is this data for assessing the effect of sulfur dioxide on the environment?

Key points

- Air is polluted by exhaust fumes from motor vehicles.
- Sulfur dioxide dissolves in rain to produce acid rain.
- Acid rain damages living tissues and some types of stones.
- Lichens can be used as indicators of the concentration of sulphur dioxide.
- Aquatic animals can be used as indicators of the acidity of freshwater.

Greenhouse gases

The Earth is warmed by radiation from the Sun and the warmed Earth produces **infra red radiation**. **Greenhouse gases** trap this radiation in the atmosphere. Without them, the Earth would be 33 °C colder and most living things would not survive.

Human activities, such as cutting down forests and increasing travel by air, are causing the concentrations of greenhouse gases in the atmosphere to rise above natural levels. As a result the Earth's atmosphere is slowly warming up. This is called **global warming**.

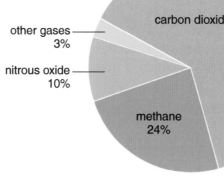

radiation from Sun passes through atmosphere and warms surface of Earth

greenhouse gases re-radiate some infra red rays back to Earth

warm Earth emits infra red radiation

Even a small rise in the Earth's temperature will cause climate change around the world. In some areas farmland will become desert. In others, flooding and storms will cause chaos.

The warming of the atmosphere is also melting the polar icecaps. This will cause sea levels to rise and some low-lying areas will be submerged.

The pie chart shows the gases that are contributing to global warming.

other gases 3%

nitrous oxide 10%

carbon dioxide 63%

methane 24%

Question

a Which gas contributes most to global warming?

CO₂ in balance

Plants remove carbon dioxide from the atmosphere during photosynthesis. Trees remove millions of tonnes of carbon dioxide from the atmosphere every year. Most of this is converted into the cells that form wood. We say that the carbon dioxide is 'locked up' in the wood.

Most living things pass carbon dioxide back to the atmosphere when they respire. These two processes have balanced each other for thousands of years, but in the last few centuries humans have interfered with this balance in two major ways.

First, there has been a massive increase in the burning of fossil fuels by industry, motor vehicles and aeroplanes. This is increasing the amount of carbon dioxide in the atmosphere. In fact it would take two trees about 100 years to remove the carbon dioxide released by a return air flight to Spain!

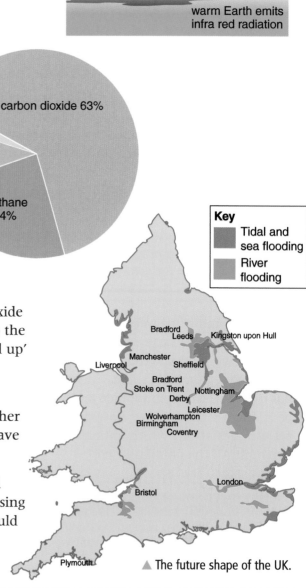

Key
- Tidal and sea flooding
- River flooding

Bradford
Leeds
Kingston upon Hull
Manchester
Liverpool
Sheffield
Bradford
Stoke on Trent
Nottingham
Derby
Leicester
Wolverhampton
Birmingham
Coventry
London
Bristol
Plymouth

▲ The future shape of the UK.

Second, large areas of forest have been cleared to produce timber and free up land for agriculture. When trees are cut down and burned, all the stored carbon dioxide is released into the atmosphere. The roots of the trees die and are decomposed by microorganisms. All microorganisms respire, so the carbon compounds in the roots are converted into carbon dioxide, which enters the atmosphere. **Deforestation**, cutting down trees, also means that the global rate of photosynthesis is reduced, so less carbon dioxide is removed from the atmosphere.

Biodiversity

Nobody knows how many species there are in the world. Scientists' estimations vary between 5 million and 80 million. Most scientists agree, however, that about half these species live in tropical rainforests. Another estimate is that deforestation is losing the world 150 species every day. Scientists call this a loss of biodiversity. Why does this matter?

One of the reasons is that drugs produced from living organisms play a big part in our daily lives. Who knows what useful chemicals there are in the species we are wiping out?

Methane – the natural greenhouse gas

Carbon dioxide is responsible for just over half of the total greenhouse effect. The other major greenhouse gas is methane.

Rice fields are under water for long periods so there is very little oxygen in the soil. The bacteria in these soils produce a lot of methane.

Cows have a four-chambered stomach. Microbes live in these chambers, helping to digest the cow's food. Because there is very little oxygen in the cow's stomach, these microbes produce methane.

As the world's population has increased, the total area of rice fields and number of cattle has risen to meet the demand for food. This has caused an increase in the amount of methane in the atmosphere.

Question

e Explain why carbon dioxide and methane are known as greenhouse gases.

Questions

b In 1990 Brazil had 543 900 000 hectares of forest. By 2000 it had lost 239 000 000 of these hectares.
(i) What proportion of forest did Brazil lose between 1990 and 2000? (Give your answer to the nearest whole number.)
(ii) What was the average yearly proportion of forest lost by Brazil between 1990 and 2000?
c Give two reasons why cutting down forests increases the amount of carbon dioxide in the atmosphere.

Question

d (i) What is meant by biodiversity?
(ii) Give one reason why it is important to protect species from extinction.

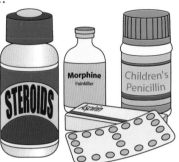

◄ All these drugs come from living organisms.

► Methane is produced by bacteria living in flooded rice fields.

Key points

- Greenhouse gases absorb energy radiated by the Earth.
- This absorbed energy may cause big changes in the Earth's climate and rises in sea levels.
- The concentration of carbon dioxide in the atmosphere is rising because of increased use of fossil fuels and increased deforestation.
- Deforestation also reduces the number of species on Earth. This is known as loss of biodiversity. Some of the species wiped out may contain substances that are useful as medicines.
- The methane concentration of the atmosphere is rising because of increased numbers of cattle and rice fields.

Increasing standard of living

In 2000 it was estimated that 50% of the tower cranes in the world were being used for building in Shanghai, China. Countries such as India and China are becoming industrialised as they bring their standard of living in line with the Western world.

All of us would like to see a better standard of living for the whole of the world's population. Improving the quality of life around the world without upsetting the Earth's natural balance is known as **sustainable development**. This means reducing energy use, conserving natural resources, conserving animal and plant species, and reducing waste and pollution. This can only be done effectively by planning. This planning needs to be done by local councils, governments and world leaders.

In this chapter, we have seen that:

- the world's population is rising faster than the supply of resources
- forests are being destroyed
- natural resources are being used up
- pollution is affecting land, sea and water
- climates are changing.

What can we, our local and national governments, do about it?

Energy use

Most energy used in industrialised countries comes from non-renewable energy resources like coal. Industrialised countries are using up non-renewable energy resources far more quickly than developing countries.

In the UK the government has set a high priority on encouraging people to use less energy. It is backing this up by introducing energy taxes. The ways in which we, as individuals, can conserve non-renewable energy resources include:

- making homes more energy efficient
- walking or using public transport instead of private cars.

▲ China catches up.

▲ Millions of people in the world live in poor conditions.

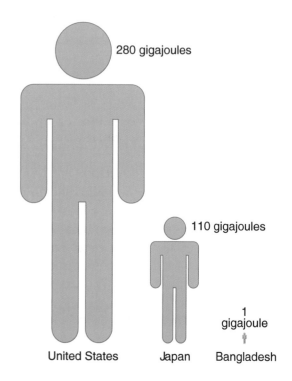

280 gigajoules

110 gigajoules

1 gigajoule

United States   Japan   Bangladesh

Question

a Look at the diagram showing energy use in different countries. Bangladesh is a developing country. Explain why each person in the USA uses so much more energy than each person in Bangladesh.

Question

b *Give four ways in which this family is reducing its demand for resources.*

Waste and recycling

This chart shows the amount of waste produced per head of population in different countries. In the UK, waste disposal is highly controlled, including taxes for using landfill sites. The government has given local councils recycling targets to meet. This means that a much higher proportion of waste is now being recycled rather than dumped. Recycling paper means that fewer forests are cut down. Recycling glass and metals means that fewer quarries are dug.

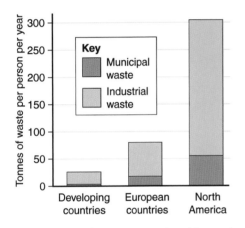

▲ Getting rid of waste is a real problem – the more people, the bigger the problem.

▲ This family has reduced its energy demands.

▶ Recycling materials helps to conserve natural resources.

Question

c *Municipal waste is the waste that is collected by refuse lorries from homes and shops. Industrial countries produce far more municipal waste per person than developing countries. Suggest reasons for this.*

One way of conserving precious natural resources is to **recycle** them. We can all contribute to this by using recycling bins. Many local councils now provide households with two bins: one for rubbish that can be recycled and one for waste that cannot be recycled.

Question

d *How does recycling each of the following save natural resources:*
(i) newspaper (ii) bottles (iii) aluminium cans?

Key points

- Sustainable development means improving quality of life without destroying resources for the future.
- Sustainable development means using natural resources and energy carefully, in planned ways.
- Natural resources can be conserved by recycling paper, glass and metals.

Investigating climate change

The photograph shows a scientist examining part of a 3 km-long ice core drilled from the Antarctic. As the ice was formed, air bubbles were trapped in it. Scientists can analyse the air from these bubbles to find the composition of air, including the concentration of carbon dioxide, trapped from hundreds of thousands of years ago to the present day. They can also estimate the temperature of the atmosphere when the bubbles were trapped.

The graphs show some of their results.

▲ Every chunk of ice core tells the story of the atmosphere it was formed in.

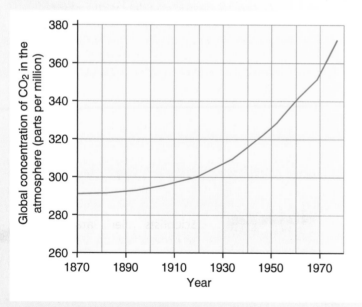

Question

a **(i)** Describe how the concentration of carbon dioxide in the atmosphere changed between 1870 and 1970.

(ii) Explain how the graphs provide evidence that carbon dioxide is a greenhouse gas.

(iii) Explain why the graphs do not prove that carbon dioxide is responsible for global warming.

Some solutions?

The world is in danger of running out of the natural resources that we have come to depend on. People in industrialised countries and in developing countries might have very different ideas about how to solve the problem.

> **Question**
>
> **b** Look at the cartoon.
> (i) What is the solution proposed by the man from the industrialised country?
> (ii) What is the solution proposed by the man from the developing country?
> (iii) What do you think is the solution to the problem?

Planning

Why aren't all governments acting immediately? Change costs money and governments that put prices up are usually not popular. We continue to use electricity from gas-fired power stations because it is cheaper than electricity from renewable resources.

But changes will have to be made and these must be planned. Some changes need to be planned by local councils. Some need to be planned by governments. Some, such as the fight against global warming, need to be tackled by world leaders working together.

The environmental pressure group Greenpeace issued this leaflet of demands:

> **Questions**
>
> **c** Which of the changes in the leaflet below left should be planned by (i) local councils, (ii) governments and (iii) world leaders?
> **d** Which of these Greenpeace demands are most and least likely to be met? Explain the reasons for your answers.

SEVEN STEPS TO SAVE THE PLANET

1 An immediate halt to all oil exploration

2 A switch from fossil fuels to wind, wave and solar energy within 40 years

3 Mass production of solar panels to make solar power more affordable

4 All new homes to be built with highest possible energy-saving standards

5 Insulation and energy efficiency increased on existing homes

6 Public transport and cycle routes improved and people encouraged to walk rather than drive

7 Campaigns to educate the public on energy efficiency

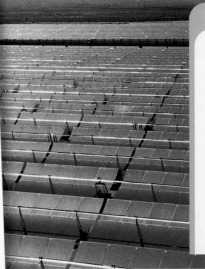

▼ In hot climates solar panels can be an effective way of generating electricity.

> **Key points**
>
> ● Scientists collect data to provide evidence for environmental change.
> ● A correlation between two variables does not always mean that one is causing the other.
> ● We need to plan future development carefully to make sure it is sustainable.
> ● Different people have different ideas about how this should be done.

1 The drawing shows a cactus. Cacti live in hot, dry places.

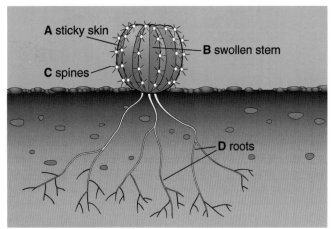

Match words **A**, **B**, **C** and **D** with the spaces **1–4** in the sentences.

Water is stored in the _____**1**_____. The _____**2**_____ are adapted to collect water over a wide area. The _____**3**_____ reduces the loss of water by evaporation. Animals are deterred from eating the cactus by the _____**4**_____.

2 A class investigated how temperature affected the number of earthworms in soil.

Each soil sample was taken from the same field and was one metre square and 15 cm deep. The air temperature was taken and earthworms were counted on the same day each month.

The results are shown in the table below.

	Jan	Feb	Mar	Apr	May	Jun	Jul	Aug	Sep	Oct	Nov	Dec
Air temp. (°C)	3	1	1	5	8	15	20	16	12	9	8	6
No. of worms	20	5	8	33	75	12	9	15	35	43	75	53

a i What were the control variables in the investigation? *(2 marks)*

ii Give **two** ways in which the investigation could have been improved to give more reliable results. *(2 marks)*

b i Plot the data on one graph. *(2 marks)*

ii In which month were the fewest earthworms found? Suggest an explanation for this. *(2 marks)*

c The class decided that another factor might also be affecting the number of earthworms.

They decided to find out the rainfall for the area. Suggest where they might find this information. *(1 mark)*

The table below shows the monthly rainfall for the area.

	Jan	Feb	Mar	Apr	May	Jun	Jul	Aug	Sep	Oct	Nov	Dec
Total rainfall (mm)	45	33	28	55	75	25	8	12	35	45	60	55

d i Add this data to the graph. *(1 mark)*

ii In which two months were most earthworms found? *(2 marks)*

iii Which conditions do earthworms prefer? *(1 mark)*

iv Outline an investigation you could do in a laboratory to find if earthworms prefer these conditions. *(2 marks)*

3 The table is about producing plants and animals with certain characteristics.

Match words from the list with each of the numbers **1–4** in the table.

A clones **B** cuttings

C enzymes **D** genes

	Function
1	used to 'cut out' genes
2	groups of genetically identical individuals
3	control inherited characteristics
4	produces genetically identical plants

4 The table is about methods of producing offspring.

Match words from the list with each of the numbers **1–4** in the table.

A asexual reproduction

B genetic engineering

C sexual reproduction

D tissue culture

	Function
1	transferring genes from one species to another
2	produces offspring with no fusion of gametes
3	producing offspring from a small group of cells
4	produces offspring with a mixture of characteristics from two parents

5 The diagram shows the technique involved in adult cell cloning.

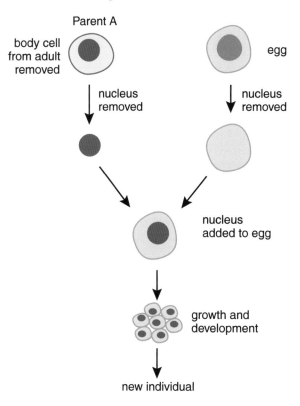

Parent A

body cell from adult removed

egg

nucleus removed

nucleus removed

nucleus added to egg

growth and development

new individual

a This technique involves
   A asexual reproduction
   B fertilisation
   C mutation
   D sexual reproduction.

b The new individual is identical to Parent A because they
   A have the same genes
   B are formed by the fusion of gametes
   C have developed in the same conditions
   D have the same enzymes.

c Some people object to producing animals in this way. This is most likely to be because cloning raises issues which are
   A economic
   B ethical
   C scientific
   D social.

d Which one of the following is **not** contained in a nucleus?
   A cells
   B chromosomes
   C DNA
   D genes

6 Plant tissue culture is a method used to produce new plants. The flow diagram shows one method of plant tissue culture.

Small piece of tissue is removed from a plant, e.g. a piece of root or stem tissue
The tissue is transferred to a culture medium
New tissue develops containing unspecialised cells
New tissue is transferred to a new culture medium
New specialised shoots and root cells develop
Developing plants are separated and grown under optimum conditions

a Name the type of reproduction involved in plant tissue culture. *(1 mark)*

b Describe **two** advantages of producing plants using this method rather than from seed. *(2 marks)*

c Explain why a disease is more likely to destroy a whole batch of plants grown by plant tissue culture than a batch of plants grown from seeds. *(2 marks)*

d Suggest **two** factors which need to be controlled to create the optimum growing conditions for the developing plants *(2 marks)*

7 The figure below shows a fossil of *Archaeopteryx*.

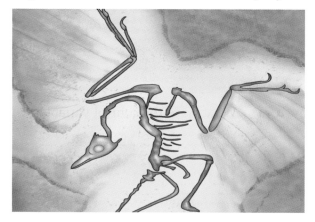

a The fossil shows that *Archaeopteryx* had jaws with teeth, and bones similar to a reptile. Scientists decided it was a bird and not a reptile. What evidence from the figure shows that *Archaeopteryx* was a bird? *(2 marks)*

b Explain why the fossil shows bones and teeth but not soft tissues such as muscle and skin.
(1 mark)

c Describe one way that fossils provide evidence of evolution. *(1 mark)*

8 The diagram shows the evolution of some groups of animals.

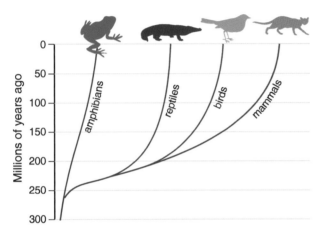

a Which group of animal, shown in the diagram, evolved first?
   A amphibians   **B** birds
   C mammals   **D** reptiles

b How many millions of years ago did the first mammals evolve?
   A 100   **B** 150   **C** 200   **D** 250

c Animals below the line for 'Present Day' are extinct. Which of the following would **not** cause animals to become extinct?
   A competition   **B** diseases
   C fossils   **D** predators

d Individuals in a species often have different characteristics. This is due to differences in their
   A cytoplasm   **B** enzymes
   C genes   **D** hormones.

9 The diagram shows a village and its surroundings.

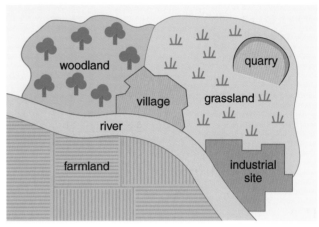

Match words **A**, **B**, **C** and **D** with the spaces **1–4** in the sentences.
   A carbon dioxide   **B** fertiliser
   C sewage   **D** sulfur dioxide

The trees remove ____1____ from the atmosphere. The atmosphere might be polluted by ____2____ from the industrial site. The river might be polluted by ____3____ from the village and by ____4____ from the farmland.

10 The graph shows damage caused by air pollution to the crown (top branches) of three species of trees in a UK forest.

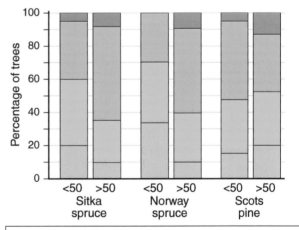

Match the numbers, **A**, **B**, **C** and **D** with the statements **1–4** in the table.

  A 10
  B 15
  C 40
  D 45

1	The percentage of Sitka spruce trees more than 50 years old with no damage to the crown.
2	The percentage of Norway spruce trees less than 50 years old with slight damage to the crown.
3	The percentage of Scots pine trees less than 50 years old with medium damage to the crown.
4	The highest percentage of tree with severe damage to the crown.

11 a Give **two** reasons why tropical rainforests are being cut down at a great rate. *(2 marks)*

  b Cutting down rainforests reduces biodiversity.
    i Explain what is meant by *biodiversity*.
      (1 mark)
    ii Give one consequence for humans of a reduction in the biodiversity of tropical forests. *(1 mark)*

  c Biodiversity can be preserved by protecting parts of rainforests.

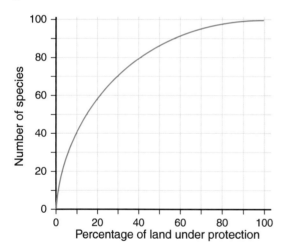

    i Describe, in as much detail as you can, the relationship between percentage of land protected and number of species preserved. *(3 marks)*
    ii Explain why small-scale protection projects can be very effective. *(1 mark)*

12 In 1952 in London there was a thick fog for several days in December. This fog trapped air pollutants. The graph shows the concentration of sulfur dioxide and smoke particles in the atmosphere. It also shows the number of deaths per day.

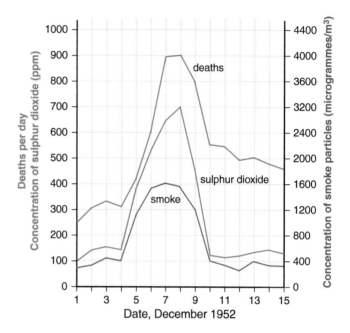

  a What was the maximum number of deaths per day?
    A 400   **B** 700   **C** 900   **D** 3600

  b What was the concentration of smoke particles on December 7th?
    A 400 micrograms per cubic decimetre
    B 650 micrograms per cubic decimetre
    C 900 micrograms per cubic decimetre
    D 1600 micrograms per cubic decimetre

  c The data
    A proves that sulfur dioxide caused the large number of deaths.
    B shows an exact correlation between sulfur dioxide concentration and the number of deaths
    C shows partial correlation between sulfur dioxide concentration and the number of deaths
    D shows that the number of deaths depends on smoke particle concentration.

  d Sulfur dioxide dissolves in water to produce
    A an acid solution   **B** an alkaline solution
    C a neutral solution   **D** sewage.

Earth provides

Caves and skyscrapers both give us shelter. Caves are just natural holes in limestone rock. Skyscrapers are built using concrete, which is made from limestone. How did we get from one to the other? Creative ideas in science and technology have helped us to take raw materials from the Earth and turn them into new building materials.

Using natural rock

We get everything we use from the Earth. Much of it comes from the rocks. Sometimes we can just take rock straight out of the ground. Blocks of stone are simply cut out of quarries. Many old buildings in cities such as Oxford were built using blocks of limestone. Limestone is a very common rock.

Science to the rescue

You can't build skyscrapers from blocks of limestone straight out of the Earth. The raw materials have to be changed before we use them. They are changed by chemical reactions.

▲ Many old buildings in Oxford are built from limestone.

Modern buildings are made from new materials like concrete and glass. Concrete and glass have been used for a long time. The Romans used concrete for buildings nearly 2000 years ago, but over the last 100 years it has become the most important building material of all.

The Romans also used glass – but it could only be made in small pieces, even as recently as Elizabethan times. Glass technology has also improved over the last 100 years. Now we can make huge sheets to cover our fantastic modern skyscrapers, such as the Lloyd's building in London.

◄▶ Glass technology has come a long way since Elizabethan times.

What are the raw materials?

Concrete is made from limestone and clay mixed with sand and gravel.

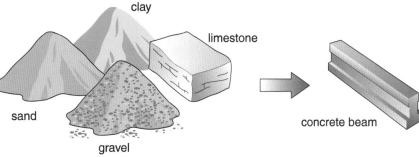

Glass is made from sand and limestone.

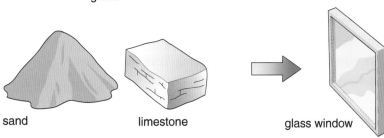

The effects of extraction

Digging out rocks leaves big holes – quarries. Unfortunately, the best limestone is often found in scenic areas, such as the Peak District. Quarries can be ugly places that spoil the natural beauty of the landscape. Quarries are also noisy and dirty. Most people wouldn't like one next to their home, but we need the limestone for making concrete and other important materials. Quarries also provide employment for people living in the area.

Think about what you will find out in this section

How can we make new building materials from the Earth?

How can we decide which new materials are best for the job?

How do we evaluate developments in building materials?

How do scientists use ideas about atoms to explain what is happening when they make new materials?

Tiny particles

Smash some limestone with a hammer and you would get lots of little bits of limestone, but limestone is made of really tiny **atoms** – 3 million million million million of them in 100 g of limestone.

▲ You won't get the atoms out this way!

> **Question**
>
> **a** Why can't you see atoms through a microscope?

Naming the atoms

Limestone is made of three different types of atoms: calcium, carbon and oxygen. These atoms make limestone when they are joined together in a chemical compound called calcium carbonate. Glass is made from calcium, silicon and oxygen atoms. These atoms make glass when they are joined together in a chemical compound called calcium silicate.

There are about 100 different kinds of atoms. Each element has a symbol. Calcium is Ca, carbon is C, oxygen is O and silicon is Si.

> **Question**
>
> **b** How many atoms are there in calcium carbonate?

Inside the atom

Atoms are made of even smaller particles. Atoms have a small **nucleus** in the middle of them. Whizzing around the nucleus are tiny **electrons**. These tiny electrons control the chemical reactions between elements and compounds.

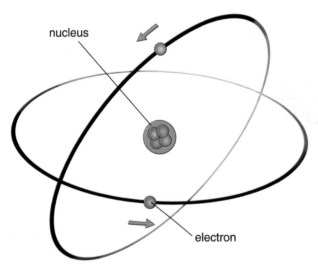

nucleus

electron

> **Question**
>
> **c** Where is the nucleus in an atom?

Electrons rule chemistry

When chemical reactions take place electrons may move from one atom to another or be shared between atoms.

Metals like to lose electrons. Non-metals like to gain them. Sodium and chlorine form a compound called sodium chloride when electrons move from sodium to chlorine. This is common salt. The sodium and chlorine atoms are stacked up in a **lattice**.

Non-metals share electrons. The atoms form molecules. Carbon and oxygen form carbon dioxide (CO_2) like this. Hydrogen and oxygen form water (H_2O).

The moving or sharing of electrons forms **chemical bonds** between the atoms in elements to make new chemical compounds.

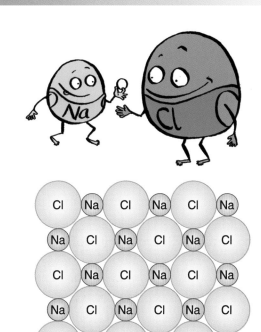

Conserving atoms

When chemicals react, the atoms just rearrange themselves. You always end up with the same number of each type of atom. You can show this as a word equation and as a balanced equation. For example, calcium oxide reacts with carbon dioxide:

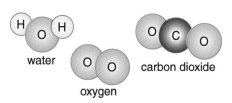

water   oxygen   carbon dioxide

$$\text{calcium oxide} + \text{carbon dioxide} \rightarrow \text{calcium carbonate}$$

$$CaO + CO_2 \rightarrow CaCO_3$$

> **Question**
>
> **d** How many of each type of atom are there on each side of this reaction?

Different properties

Different elements have different properties. For example, calcium is a metal, carbon is a black solid non-metal and oxygen is a gas. The different elements are arranged in the periodic table. Similar elements are found in the same group. Compounds have completely different properties from the elements that make them. Calcium carbonate is limestone.

> **Question**
>
> **e** Coal contains carbon atoms. Why isn't it like limestone?

Key points

- Atoms have a small nucleus, around which are electrons.
- Atoms and symbols are used to represent and explain what is happening in chemical reactions.
- Elements form compounds by giving and taking electrons or by sharing electrons to form chemical bonds.

Using limestone

Limestone is a common rock with many uses. It is found in many beautiful parts of the country and dug out of large quarries. Some people say that these quarries spoil the beauty of the countryside.

Limestone has been used as a building stone for thousands of years. Today, most limestone is used as a raw material to make something new. It is changed by chemical reactions into other useful products, such as cement, or broken up for roads or concrete.

Limestone is a very important raw material. About 90 million tonnes of limestone are quarried every year in Britain alone. This is used:

- for buildings or road 'chippings' (66 million tonnes)
- to get rid of acid in lakes or soils (1 million tonnes)
- to make cement or glass (23 million tonnes).

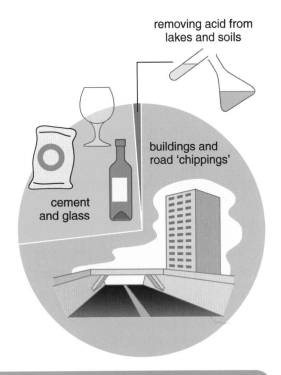

removing acid from lakes and soils

buildings and road 'chippings'

cement and glass

Questions

a *Limestone is very useful – but should it be quarried in countryside like this? List two possible problems.*

b *Approximately how much of all quarried limestone is used for cement and glass…$\frac{1}{4}$, $\frac{1}{3}$ or $\frac{1}{2}$ of it?*

Making new materials from limestone

Limestone is the compound calcium carbonate. Its chemical formula is $CaCO_3$.

Question

c For each calcium atom in calcium carbonate how many carbon and oxygen atoms are there?

carbon dioxide gas

solid calcium carbonate

solid calcium oxide

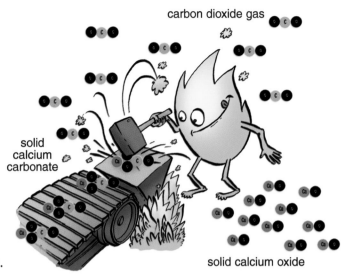

If limestone is heated strongly, the compound breaks down.

- The carbon atom takes two oxygen atoms to form carbon dioxide.
- The calcium atom is left with just one oxygen atom. This is calcium oxide, which is also called **quicklime**.

$$\text{calcium carbonate} \xrightarrow{\text{heat}} \text{calcium oxide} + \text{carbon dioxide}$$
$$CaCO_3 \longrightarrow CaO + CO_2$$

Question

d What will happen to the carbon dioxide in this reaction?

Breaking a compound by heating it like this is called **thermal decomposition**. As in all chemical reactions you have the same atoms – they have just been rearranged. So the products have the same mass as the reactants. Mass is always conserved in chemical reactions.

Many other metal carbonates break down like this, for example copper carbonate:

$$\text{copper carbonate} \xrightarrow{\text{thermal decomposition}} \text{copper oxide} + \text{carbon dioxide}$$
$$CuCO_3 \longrightarrow CuO + CO_2$$

Question

e If you started with 100 g of limestone, what would be the mass of calcium oxide and carbon dioxide after a thermal decomposition reaction?

Question

f What two compounds would you get if you heated iron carbonate?

Reacting quicklime

Quicklime is a very strong alkali. It reacts with water, forming calcium hydroxide, which is also called **slaked lime**.

$$\text{calcium oxide} + \text{water} \rightarrow \text{calcium hydroxide}$$
$$CaO + H_2O \rightarrow Ca(OH)_2$$

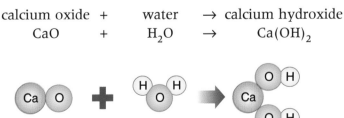

Key points

- The formula of a compound shows the number and type of atoms in it.
- In chemical reactions, atoms are rearranged. No atoms are lost or made, so we can represent reactions using balanced equations.
- Limestone can be used as a building material or as a raw material for new products.
- Limestone ($CaCO_3$) and other carbonates can be broken down by thermal decomposition.

Rock or not?

These statues look very similar. One took a sculptor many weeks to make by shaping the limestone. The other was made by simply pouring concrete into a mould. Concrete is like a rock, but it isn't.

Making mortar, cement and concrete

Houses in Britain used to be built from bricks stuck together with **mortar**. The mortar was made by mixing slaked lime (made from quicklime) with water to make a thick paste. This dried and set hard between the bricks, holding them together.

> **Question**
>
> **a** Which rock is used to make mortar?

Today, most quarried limestone is used to make **cement**. The limestone is heated with clay in a big oven called a kiln. The product is then ground to form a light grey powder. This is cement, which can be used to stick bricks together like mortar.

> **Question**
>
> **b** Bricks are made from clay. What else is clay used for in modern houses?

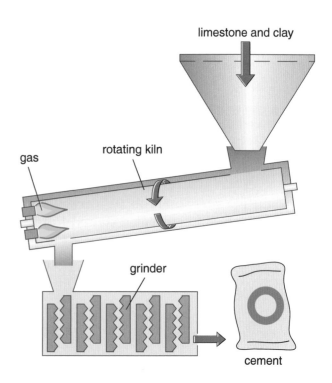

Most cement is mixed with sand, gravel and water to make **concrete**. This is cheaper and stronger than pure cement. Freshly mixed concrete forms a thick liquid, which can be poured into any shape. Slow chemical reactions make it set after a few hours and it becomes rock hard. It is used to make roads, bridges and buildings.

▶ Pouring cement – it will soon set rock hard!

Making glass

Glass is a very important building material. Many modern buildings are completely covered in glass. You get fantastic views out from the huge windows, and plenty of natural light gets in so you save on electricity bills. Glass can also look fantastic. Concrete on its own can seem dull and boring by comparison.

Glass is also made from limestone. The limestone reacts with another common raw material – sand.

Pure sand is usually grains of silicon dioxide. Glass is made by heating very pure sand with limestone and a little soda (sodium carbonate). The sand and limestone react, melt and carbon dioxide is released. When this liquid cools, it forms a hard, but brittle, transparent material – glass.

▲ Making large glass panels.

limestone   +   sand   →   glass   +   waste gas

calcium carbonate + silicon dioxide → calcium silicate + carbon dioxide

Questions

c Copy and complete this equation. Make sure it is balanced.
$CaCO_3 + SiO_2 \rightarrow CaSiO_3 +$ _____
d Suggest one possible disadvantage of a glass-fronted building on a very sunny day.

Key points

● Limestone can be used as the raw material for quicklime, slaked lime, cement and concrete.
● Limestone is also the raw material for glass, along with sand.

Concrete: beauty or the beast?

The Royal Festival Hall in London is made of concrete. The concrete is exposed for all to see. Some people like it, but many find the building ugly. The concrete gets streaked and dirty very quickly.

The beautiful Bahai Lotus temple in Delhi is also made of concrete. You can build whatever you like from concrete. There are many imaginative concrete buildings around the world.

▲ The Royal Festival Hall was built in 1951 as part of the Festival of Britain.

▲ The Bahai Lotus temple in Delhi, India, which was built in 1980.

Getting the best out of concrete

Concrete is such a fantastic 'artificial rock' that it is much better than the real thing in many ways.

- Powdered cement is easy to store and transport.
- Sand and gravel can be found almost anywhere and are cheap materials.
- Liquid concrete can be easily mixed on site.
- Concrete can be moulded into any shape you want.
- Once concrete sets it really is rock hard!

Concrete and quarried limestone have similar properties. Both are very strong, so they can support very large buildings such as skyscrapers, but if you stretch or bend them, they can crack easily. Developments in concrete technology have helped to overcome this problem. Concrete beams or girders are reinforced with steel rods. These stop the concrete from stretching and cracking. It isn't possible to reinforce limestone in this way.

> **Question**
>
> **a** Give three reasons why it is easier to use concrete for building than solid limestone.

> **Question**
>
> **b** Skyscrapers sway and bend slightly. Why does their concrete structure have to be reinforced with steel rods?

Getting the best out of glass

Glass is a wonderful material. Glass-fronted offices give a feeling of space because you can see outside. Many people like to feel the warmth of the sun through the glass, and this can help to cut down heating costs as well. This can be a problem in the summer, when the building could trap heat like a greenhouse and become too hot to work in.

Thanks to technology, glass can be made much tougher. Toughened glass is five times as strong as ordinary glass. This toughened glass can be used where ordinary glass would shatter, for example in large glass panels for doors, large shop windows, fish tanks, car windscreens and facing panels for skyscrapers.

Car windscreen glass still shatters in a crash, but it breaks into small but chunky pieces that are not as sharp as ordinary glass slivers. More modern windscreens have an internal plastic layer that holds all the pieces together. As technology improves, glass gets better and better. However, even the best glass will shatter if there is an explosion or a building is shaken by a powerful earthquake.

Glass-fronted buildings can be good to work in as they are light and airy, but they can overheat in summer if not designed well. Broken glass is also very sharp and dangerous.

Questions

c *Glass is a poor insulator. What problem might there be working in a glass-fronted office in winter?*
d *Ordinary glass has one problematic property. Suggest what this is.*

Question

e *Why is glass in a door more likely to shatter than glass in a window?*

Question

f *Why would it be a bad idea to stand at the base of a skyscraper during a powerful earthquake?*

Key points

- Concrete is a strong, inexpensive and easy to use material for building.
- Concrete is very versatile but needs to be reinforced with steel to stop it cracking if used for beams or girders.
- Some people think concrete buildings are ugly.

Useful metals

We use metals to make everything from aeroplanes to tin cans and electrical wiring. We use such a lot of metals because they have many useful properties.

What's so good about metals?

The properties of metals make them useful.

- Metals are strong: we use them to built great machines or structures.
- Metals are hard: the cogs and gears in our machines must not wear away.
- Metals are easy to shape: car body parts are just pressed out of sheet metal.
- Metals have high melting points: engines have to withstand high temperatures.
- Metals conduct electricity: none of our technology could work without electrical wiring.
- Metals conduct heat: not just useful for frying eggs. Computer chips have metal 'heat sinks' to take away heat energy and stop them overheating.

Are there problems with metals?

Some properties of metals are less useful.

- Some metals are rare or even very rare: gold would have many more uses if it wasn't so rare and therefore so expensive.
- Metals can be hard to extract from rocks: aluminium is twice as common as iron, but costs six times as much as it is so hard to extract.
- Pure metals are too soft to be of much use: scientists have worked out how to make alloys with improved strength.
- Metals corrode. We spend money and effort getting iron from rocks, but if you leave any iron object lying about in the rain you end up with a pile of rust.

◄ All that effort to end up with a pile of rust...

And what if we run out?

Modern technology relies on metals but at present rates, using today's mining techniques, we risk running out of many important metals over the next 100 years. What can we do to stop this happening? Some scientists are trying to find new sources of metals and new ways of extracting them. Others are looking for better ways to recycle the metals we have.

What's the best material?

In China, builders use bamboo scaffolding rather than the steel scaffolding we use here. Let's look at why. To evaluate the best materials, we have to look at their properties, their costs and their availability.

Bamboo is a completely natural, renewable raw material grown locally. Steel has to be manufactured and China is currently having to import large amounts of it to keep up with the demand. Bamboo is much cheaper than steel. Bamboo is also lighter for workers to handle and move around, yet it is strong enough to use as scaffolding, even on tall buildings. And when you have finished with it, it can be left to rot naturally without harming the environment.

Think about what you will find out in this section

How do we get metals from their ores?

How are iron and steel made?

How do we improve metals to make them more useful to us?

How can science help to make recycling easier?

How can we evaluate our use of metals for structural and smart materials?

Getting at the metal

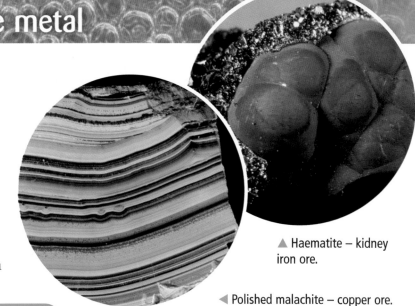

▲ Haematite – kidney iron ore.

◄ Polished malachite – copper ore.

Where do metals come from?

Metals are found in rocks, but rarely as pure metal. They are usually chemically combined with other elements in chemical compounds. Some metals are quite common. A field of mud contains tonnes of aluminium. The trouble is it would cost too much to get the metal out. Some rocks contain metal that is easier and cheaper to extract. Rocks like this are called **ores**. Iron ore and copper ore are two good examples.

Question

a Why do you think it is important that it is cheap and easy to extract metals from ores?

Which metals are common?

The table shows some metals we use a lot. If you look at the table you can see that some of them are very common but some (copper, tin and gold) are really very rare.

Metal	Percentage of the rocks of the Earth's crust
aluminium	7
iron	4
copper	0.0045
tin	0.0002
gold	0.0000005

Question

b Which of the metals in the table do you think we use most?

▼ Gold was well known to the ancient Egyptians.

Of these metals, iron is the one we use the most. About 1 billion tonnes of iron is produced every year. That's 20 times as much iron as all the other metals put together! Production on a large scale helps to make iron cheap compared to aluminium.

Metal	Price per tonne (spring 2005)
aluminium	£1800
copper	£3300
iron	£300

Question

c Copper is easier to get from its ore than iron. Why do you think it is so much more expensive? (Hint: look at the first table on this spread).

◀ There is plenty of aluminium in this mud.

Reactivity and metals

Some metals are found in a pure form. Gold is a very unreactive metal. It does not react with other elements. It is very rare but, if you are very lucky, you could find a nugget of pure gold!

Most metals are found as compounds. Aluminium is a very reactive metal. Because of this, aluminium atoms form strong compounds with other atoms, such as oxygen. Mud contains plenty of aluminium, but you cannot easily get it out. You will never find a lump of pure aluminium in rocks.

most reactive

aluminium

carbon

iron

tin

copper

gold

least reactive

Question

d What properties of gold would make it stand out in a pile of gravel?

Using reactivity to extract metals

The reactivity series leads us to a way to extract metals from ores. Carbon is more reactive than some metals so it can displace these metals from their ores. This is a method that is used wherever possible for getting a metal from its ore. It works for iron. The reaction is:

iron oxide + carbon → iron + carbon dioxide

It works for copper, too.

copper oxide + carbon → copper + carbon dioxide
$$2CuO + C \rightarrow 2Cu + CO_2$$

Key points

- Ores are found in the Earth and contain metal compounds from which the metal can be extracted.
- Some metals are more economic to extract than others.

Question

e Which other metals can be displaced from their ores by carbon in this way?

Wanted by all

Iron is the most widely used metal so we need to produce plenty of it to supply the world. Over the last 300 years, scientists have refined the carbon reduction method to make it very efficient. Iron is not as easy to extract as copper or tin, which are less reactive.

Iron plays hard to get

Using carbon to extract copper needs a kick-start of energy to get the reaction going. For copper oxide, the heat from a simple fire will do it but for iron oxide much more energy is needed to get the reaction started. You need a special furnace called a blast furnace to get iron from its ore.

Iron ore (iron oxide, Fe_2O_3), coke (carbon) and limestone are tipped in at the top of the blast furnace. Inside the blast furnace the temperature is hot enough for the oxygen to separate from the iron in the iron oxide ore and produce iron and carbon dioxide. The iron melts, sinks down through the furnace and can be removed. The limestone helps to remove impurities from the iron.

This reaction, where oxygen is removed from a metal oxide, is called a **reduction** reaction:

iron oxide + carbon → iron + carbon dioxide

$$2Fe_2O_3 + 3C \rightarrow 4Fe + 3CO_2$$

▲ Pouring the purified molten iron from a blast furnace.

Questions

a Give two reasons why the blast furnace needs to run at such a high temperature to make iron. (Hint: for reason two: copper melts at 1080 °C, iron melts at 1540 °C.)

b Tin can be made from tin oxide as follows:

tin oxide + carbon → tin + carbon dioxide

$$SnO_2 + C \rightarrow Sn + _____$$

(i) What is being reduced in this reaction?

(ii) Complete the balanced chemical equation for this reaction.

Cast iron

The iron that comes from a blast furnace is called cast iron. It contains 96% iron and is very hard but quite brittle. Over 100 years ago this was used to make everything from bridge girders to manhole covers to bathtubs, but today it is thought to be too brittle for most uses.

Carbon makes the difference

Cast iron is not pure iron as it contains 4% carbon. It is not too hard to get rid of this, but while 96% iron is too brittle, 100% pure iron is too soft to use for construction. It can be made much harder by leaving a little carbon – just 1% or so. This new material is hard and tough! We call this new material steel. Steel is an **alloy** of iron. Most iron is converted to steel these days because it is more useful to us.

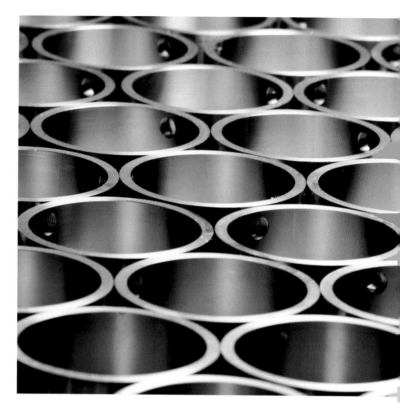

> **Question**
>
> **c** Do you think cars are built from cast iron or steel?

The price of steel

The cost of metal changes over time. It depends on what it costs to extract the metal from its ore, its availability and the demand for it. These factors are affected by both changes in technology and changes in global economics and politics.

This graph shows how the world price of steel has changed over the last 50 years. The prices have been adjusted to match today's values. Iron ore is very common so the supply and price of iron ore has not affected the price very much.

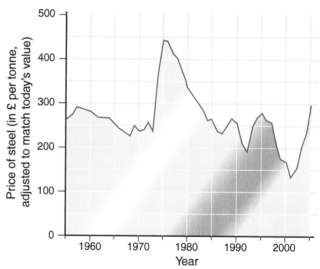

> **Questions**
>
> **d** When did steel cost the most (when adjusted to match today's values)?
> **e** A lot of steel-making firms went out of business around 1990. Suggest a possible reason for this.
> **f** Over the last few years, China has started to import a lot of steel. What effect has this had on the price?

> **Key points**
>
> - Iron is extracted from its ore in a blast furnace.
> - Impurities make cast iron very brittle but pure iron is very soft.
> - Iron is most useful when converted into steel with the addition of 1% carbon, making an alloy.
> - Whether it is worth mining an ore or not depends on the price that you can sell the metal for, and that changes from year to year.

Why are alloys harder than pure metal?

In a pure metal all the atoms are the same, stacked up in a regular way. The layers of atoms can slide over one another. In an alloy, a few different atoms have been added. These are not the same size so they mess up the pattern. The layers jam up and can't slide so easily. This makes the alloy harder.

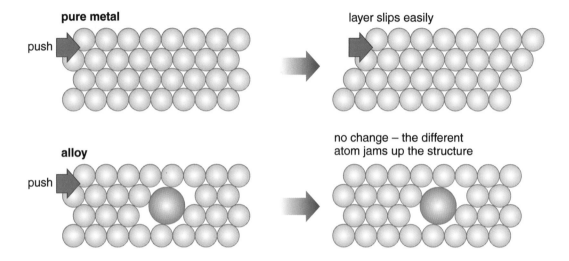

The best steel for the job

Adding carbon to iron makes it harder as the atom layers can't slide so easily, but adding too much carbon makes it weaker. It becomes brittle. The graph shows how adding carbon changes steel.

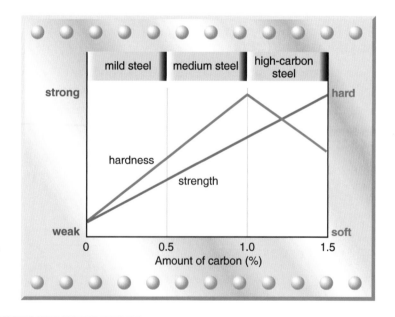

- Mild steel is soft and thin sheets can easily be shaped for car bodies.
- Medium steel is harder and is strong enough for hammers.
- High-carbon steel is hard enough to give a good cutting edge for scissors and drill bits. It is not as strong as medium steel and can be brittle.

Questions

a What percentage of carbon gives the strongest steel?
b Why do drill bits snap easily?
c Roger wants to make tools by melting old car bodies and pouring the steel into moulds. What would he need to add to his molten iron for this to work well?

Special steels that don't corrode

Sometimes other metals are added to steel to give new alloys with special properties. Chromium is added to make stainless steel, which does not rust: perfect for cutlery – or expensive cars.

▶ No rust problems for this stainless steel car!

Questions

d Suggest other uses for stainless steel, where it is important that the metal does not corrode.

e Suggest a reason why all cars are not made from stainless steel.

Smart alloys

Pure copper, aluminium and even gold would be too soft to use on their own. They are mixed with other metals to make useful alloys, which are harder.

Gold is mixed with a little copper for jewellery.

Aluminium is mixed with magnesium and a little copper to build aeroplanes.

Metal scientists can now make fantastic new alloys with improved properties. Special smart alloys can be bent at low temperatures – but they snap back to shape if they warm up.

Some people get ill because of collapsed arteries. A wire grid can keep the arteries open, but how can it be put in place? Using a smart alloy the grid can be cooled and squashed to fit in easily. It snaps back to size as it warms up and opens the artery.

▲ Never mind… it'll be as good as new in the summer!!

Question

f Why will these 'smart cars' be fixed in the summer?

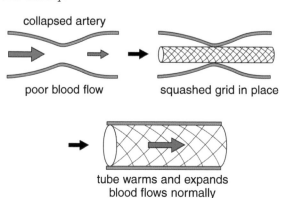

collapsed artery

poor blood flow

squashed grid in place

tube warms and expands blood flows normally

Key points

- Alloys are mixtures of metals or a metal and carbon, which are harder due to the different sized atoms in the layered structure.
- There are many different types of steel alloys with different properties depending on the carbon content.
- There are many other alloys made with other metals to give useful properties.
- Smart alloys can return to their original shape after being deformed, which can be a useful property, but such alloys are expensive to produce.

Rare but very useful

What comes out of the biggest hole in the world? The answer is copper! Copper is a very important metal, but it is much rarer than iron so when mining engineers find concentrated copper ore they just keep on digging until they've got it all out. The Bingham Canyon mine in Utah, USA, is now nearly 1 km deep!

Question

a *What is copper used for in the home?*

What's so good about copper?

If we look at the periodic table there is a block between Groups 2 and 3 that contains the 'everyday' metals. They are called **transition metals** and are useful as they can be bent or hammered into shape. Like all metals, they conduct heat and electricity. Transition metals like iron are very useful for building large structures: buildings, ships and planes. Copper is one of the transition metals from the periodic table.

Transition metals like copper are also very useful.

- Copper is a very good electrical conductor. It is also soft and bendy, so it's great for electrical wires.
- Copper does not corrode. This makes it useful for copper pipes for plumbing.

Question

b *Which property also makes copper useful for roofs?*

					Group 3	
					B	
					Al	
Fe	Co	Ni	Cu	Zn	Ga	Ge
Ru	Rh	Pd	Ag	Cd	In	Sn
Os	Ir	Pt	Au	Hg	Tl	Pb

▼ The Bingham Canyon copper mine, Utah, USA.

Getting the metal out

Digging copper ore out of the ground is just the start of the extraction process. Copper ore may be just 1% copper. The rest is waste rock. These days acid is sprayed onto the rock to dissolve the copper into a solution. The copper is then extracted from the solution by **electrolysis** – separating the copper from the solution using electricity.

▲ 'Peacock' copper ore.

> **Question**
>
> **c** If you have copper ore with 10% or more of copper in it, it is cheap and easy to get the copper out by reduction. Why is that method not used very often today with copper ore containing just 1% copper?

But it's running out!

There are just a few really big copper mines and even these have very little copper in the rock. We use large amounts of copper and it will soon run out. What will we do then? Many mines are already reworking their old waste tips to get more of the copper out.

> **Question**
>
> **d** Make a list of all the things in the picture that rely on copper to work.

So what's the future?

Acid can get small amounts of copper from rocks but scientists are working hard to find ways of extracting even more. One new method involves using special bacteria that 'eat' the copper from the rock!

Another problem is that environmentalists object to mining companies making huge holes in the Earth. Open-cast mines like this are very large and very ugly (see opposite).

A new way of mining copper is simply to drill down to the ore. Acid is then pumped down to the ore and comes back up rich in copper.

> **Question**
>
> **e** List three reasons why you would not like to have a big copper mine near to where you live.

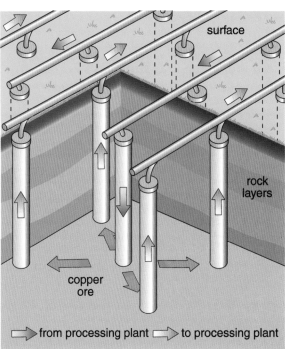

surface

rock layers

copper ore

⟹ from processing plant ⟹ to processing plant

> **Key points**
>
> ● Transition metals are good conductors of electricity and heat and can be bent into shape – properties which make them very useful.
> ● Copper is extracted by electrolysis but other ways are being developed to use ores with low copper content and limit environmental impact.

Useful metals with low densities

Steel may be fantastic for cars, trains and ships but it is very heavy for its size. An aeroplane made from steel would weigh twice as much as a modern plane. It would never get off the ground!

Aluminium is not as strong as steel but it has a very low density. This combination is just right to make a plane that is both strong enough and light enough to fly.

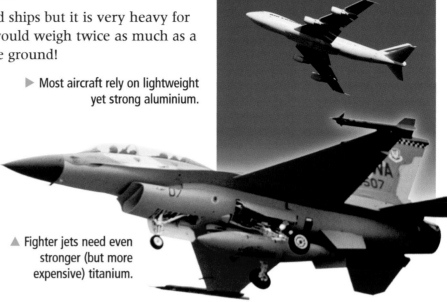

▶ Most aircraft rely on lightweight yet strong aluminium.

▲ Fighter jets need even stronger (but more expensive) titanium.

Question

a The very first aeroplanes were made out of wood and canvas so they were light enough. What do you think their big problem was?

Metal	Density (grammes per cubic centimetre)	Strength	Melting point (°C)
pure aluminium	2.7	low	660
steel	7.7	high	1540
titanium	4.5	high	1670

Supersonic fighter jets need to fly so fast that their wings would get very hot. They also have to be light yet strong to withstand enormous forces. Aluminium couldn't cope. It would be ripped apart – if it didn't melt first!

A new 'supermetal' that fighter jet manufacturers turned to was titanium, which is as strong as steel but much lighter. It also has a very high melting point – higher than steel.

Questions

b Look at the data in the table and suggest a reason why titanium might be useful for parts of fighter jets.
c The space shuttle wing edge reaches 2000 °C on re-entry. Why can't this be made of titanium?

The cost of extraction

Aluminium is more reactive than carbon. You can't get aluminium from its ore by the carbon reduction method. Aluminium ore has to be melted and then split apart using large amounts of electricity. This is a very costly process using a lot of energy and with several stages, which makes aluminium much more expensive than iron.

Question

d Aluminium cars would be lighter and so would use less fuel. Why aren't ordinary cars made from aluminium?

Titanium is also more reactive than carbon. Like aluminium, it can't be made by carbon reduction. It is much harder to extract. This means that titanium is much more expensive even than aluminium. It is only used if its superstrong properties are needed.

Resisting corrosion

Aluminium and titanium share another useful property: they do not corrode easily. That's why aluminium cooking foil stays shiny. Titanium resists corrosion much better than aluminium and stainless steel so titanium can be used safely in places where even stainless steel would corrode away, for example nuclear power stations – or inside the human body!

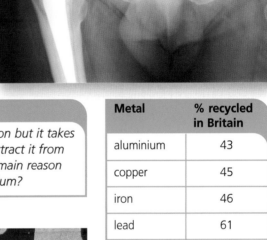

> **Question**
>
> **e** What would happen to a piece of steel left inside a human body?

Recycling metals

In Britain, we **recycle** 43% of the aluminium we use.

The table shows which metals we recycle in Britain.

> **Question**
>
> **f** Aluminium is common but it takes a lot of energy to extract it from its ore. What is the main reason for recycling aluminium?

Metal	% recycled in Britain
aluminium	43
copper	45
iron	46
lead	61
tin	30
zinc	14

> **Key points**
>
> - Aluminium and titanium have two very useful properties: low density and resistance to corrosion. However, their extraction is expensive as it requires a lot of energy.
> - We need to recycle metals as much as possible because the amount of ore is limited and extraction is costly in terms of energy and its impact on the environment.

Re-mining

Earth is like a spaceship. We all live here and we get everything we need from it. There are 6 billion people on the planet and we are consuming the Earth's resources at an alarming rate. Fossil fuels like oil and gas will soon run out. Some metal ores will become scarce in the next 100 years or so.

Metal	Proven reserves will last until ...
tin	2030
copper	2030
aluminium	2050
nickel	2100

Question

a   Which metals might run out in your lifetime?

Some people think that using up metal ore resources is not really a problem. As our technology improves we can get metals from poorer ores. Old waste tips are often 're-mined' using newer technology. Also, if a metal starts to run out, the price goes up. This means that mines that had to close when prices were low can sometimes re-open and produce metal at an economic price.

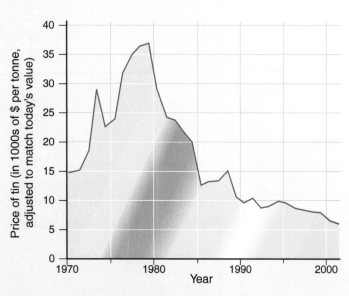

Question

b   The graph shows how the price of tin has changed over the last 40 years.
(i) Why did many old tin mines re-open in the early 1970s?
(ii) What do you think happened to many of those mines in the 1980s?

Are we exploiting poorer countries?

> It's a disgrace. We wasted all our own resources, so now we're just plundering other poorer countries.

> In the Bronze Age there were huge copper mines in North Wales. We used to export our copper then but now it's all used up so we have to import from Zambia or Chile. There's nothing wrong with that.

> But we get our copper cheap because the workers are paid such low wages. That's hardly fair, is it?

> At least the workers get a wage and can feed their families. That's better than having no job and starving.

> Until they get injured or killed! They don't have all the health and safety laws we have here. It's just exploitation!

The economics of recycling

Recycling costs money. One of the problems is that waste material may well have lots of different metals mixed up. Recycling is only economically viable at the moment for metals that are easy to separate, such as iron and steel, or rare but expensive metals, such as gold. As technology improves, perhaps we will be able to recycle all metals.

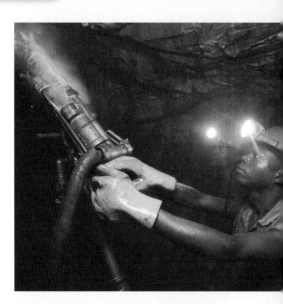

Questions

c What special property of iron and steel makes it easy to remove these metal scraps from waste material containing other metals such as copper or aluminium?

d You need to use £100 worth of energy to make 1 tonne of aluminium from its ore. It only costs £5 worth of energy to recycle 1 tonne of aluminium. How much 'energy money' would you save per tonne if you used 50% recycled aluminium and 50% new aluminium instead of 100% new aluminium?

Some scrap metal companies have discovered that it makes economic sense to ship their scrap metal around the world to countries such as India. There they can use cheap labour to hand-pick and sort the scrap into different metals. The sorted scrap is then shipped home for recycling.

Question

e Is this just good business or is it exploitation? List as many arguments as you can for and against this. Hold group discussions to pool your ideas.

Key points

- Mining, extracting and recycling metals have economic, social and environmental impacts.
- The supply of metals on the Earth is finite so we need to develop ways of using and recycling them effectively.

Oil: black gold

Our shiny new technological world relies on oil. We need oil to fuel our cars, machines and power stations. Oil helps the economy grow so we can afford new buildings and transport systems.

▲ Oil money has built gleaming cities in the deserts of the Middle East.

▲ But oil can do great damage if handled badly…

Environmental impact

There is a price to pay for using oil. The photo shows the terrible pollution caused by burning oil wells after the first Gulf War. Pollution is a problem whenever we burn fossil fuels. Oil can cause a variety of pollution problems.

- Oil slicks from wrecked tankers carrying oil kill seabirds and sealife.

- Burning oil can lead to acid rain.

- Burning oil makes carbon dioxide. This may cause global warming.

Human impact

Around 50 years ago city streets had just a few cars parked. Today you can barely see the road for cars. Almost everybody wants their own car. They want the freedom to go wherever they want, whenever they want, but look up from those same streets on a hot summer day and you will see a brown haze. This is caused by the exhaust fumes from all these cars.

Around 50 years ago asthma was uncommon. Today 15% or so of children suffer from asthma and need to use chemical inhalers to help them breathe when they have an attack. Attacks happen more often on high pollution days. Scientists monitor air quality and issue 'high pollution' warnings to help asthma sufferers.

The rise in asthma seems to match the rise in air pollution. Maybe there is a link between air pollution and asthma. Research suggests that pollen and even thunderstorms can also trigger asthma attacks, but some people think air pollution may trigger the problem in the first place.

Economic impact

North Sea oil has been great for Britain. It has helped to keep our economy healthy and created many new jobs but not all countries have been so lucky. Nigeria is the biggest oil-producing country in Africa but despite producing 2 million barrels of oil a day for the last 30 years, the standard of living for ordinary Nigerians has been falling steadily. Most of the oil money has gone to a few rich people. Meanwhile other industries have suffered and jobs have been lost. What will happen now that their oil has started to run out?

Think about what you will find out in this section

What is crude oil made of?

How can we make useful products from crude oil?

What environmental problems are linked to the use of crude oil?

How can science help to overcome any pollution problems caused by oil?

How can science help to develop new energy sources for the future, when oil runs out?

How did crude oil form?

When plants or animals die they start to rot. Eventually, if they are buried, oil and natural gas are formed. The oil and gas trapped in rocks has formed in this way over millions of years.

Question

a Oil is an ancient biomass. What do you think that means? (Hint: what does the bio bit mean in biology?)

What is crude oil made of?

Crude oil is a mixture of many different compounds. Mixtures consist of compounds or elements that are not chemically combined. Each compound in oil has its own properties. Many are very useful but mixed up together they are useless – a nasty, gooey black liquid. The different compounds must be separated out before they can be used.

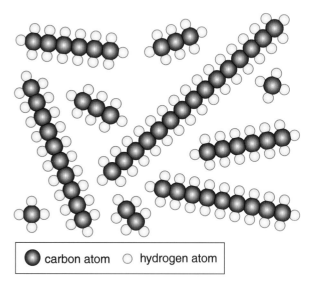

● carbon atom　○ hydrogen atom

Question

b What do you think would happen if you put crude oil into a car instead of petrol?

What kind of compounds?

Most of the compounds in crude oil are made from two types of atom only: hydrogen and carbon. There are chemical bonds between the carbon and hydrogen atoms. Compounds like this are called **hydrocarbons**.

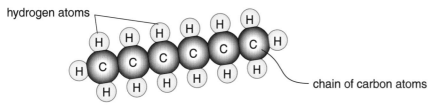

hydrogen atoms

chain of carbon atoms

There is another way we can draw hydrocarbon molecules.

$$H-\overset{\displaystyle H}{\underset{\displaystyle H}{C}}-\overset{\displaystyle H}{\underset{\displaystyle H}{C}}-\overset{\displaystyle H}{\underset{\displaystyle H}{C}}-\overset{\displaystyle H}{\underset{\displaystyle H}{C}}-\overset{\displaystyle H}{\underset{\displaystyle H}{C}}-\overset{\displaystyle H}{\underset{\displaystyle H}{C}}-H$$

Questions

c What do the 'lines' show in this drawing of a molecule?
d Count the lines from each carbon atom. How many are there?

Each carbon atom can have four chemical bonds. When all four chemical bonds are used to join carbon to carbon or carbon to hydrogen, the compound has as many hydrogen atoms in it as it can get. We say it is 'full' of hydrogen. Another word for full is 'saturated'. Your clothes get saturated with water (full of water) if you stand out in the rain. We call these compounds **saturated** hydrocarbons.

$$H-\underset{\underset{H}{|}}{\overset{\overset{H}{|}}{C}}-\underset{\underset{H}{|}}{\overset{\overset{H}{|}}{C}}-H \quad C_2H_6$$

$$H-\underset{\underset{H}{|}}{\overset{\overset{H}{|}}{C}}-\underset{\underset{H}{|}}{\overset{\overset{H}{|}}{C}}-\underset{\underset{H}{|}}{\overset{\overset{H}{|}}{C}}-H \quad C_3H_8$$

$$H-\underset{\underset{H}{|}}{\overset{\overset{H}{|}}{C}}-\underset{\underset{H}{|}}{\overset{\overset{H}{|}}{C}}-\underset{\underset{H}{|}}{\overset{\overset{H}{|}}{C}}-\underset{\underset{H}{|}}{\overset{\overset{H}{|}}{C}}-H \quad C_4H_{10}$$

$$H-\underset{\underset{H}{|}}{\overset{\overset{H}{|}}{C}}-\underset{\underset{H}{|}}{\overset{\overset{H}{|}}{C}}-\underset{\underset{H}{|}}{\overset{\overset{H}{|}}{C}}-\underset{\underset{H}{|}}{\overset{\overset{H}{|}}{C}}-\underset{\underset{H}{|}}{\overset{\overset{H}{|}}{C}}-H \quad C_5H_{12}$$

There is a pattern here. As the number of carbon atoms goes up, so does the number of hydrogen atoms. Each carbon atom in the chain can hold two hydrogen atoms. The carbon atoms at each end of the chain can hold an extra one at each end. So to find the number of hydrogen atoms in any hydrocarbon like this, simply double the number of carbon atoms and add 2. Hydrocarbons that follow this simple pattern are called the family of **alkanes**.

Long chain or short chain

Hydrocarbons have molecules of different sizes. Carbon atoms can form chains. Big molecules have long chains, small molecules have short chains.

The properties of hydrocarbons change gradually as the carbon chains get longer.

	Boiling point	
Short chains	Molecules move fast enough to boil very easily. They have low boiling points.	
Long chains	Molecules are harder to get moving. They have high boiling points.	

> **Question**
>
> **e** How many hydrogen atoms are there in a hydrocarbon molecule with seven carbon atoms?

> **Key points**
>
> - Crude oil is a mixture of different compounds. Most of these compounds are saturated hydrocarbons called alkanes.
> - C_2H_6 is an example of an alkane. Double the number of carbon atoms and add two to find the number of hydrogen atoms. The general formula is C_nH_{2n+2}
> - Different sized hydrocarbon molecules have different numbers of carbon atoms joined in a chain.
> - Short chains have low boiling points and long chains have high boiling points.

> **Question**
>
> **f** How might this help us to separate the compounds?

Boiling hydrocarbons

Crude oil is a mixture of short-chain and long-chain hydrocarbons that boil at different temperatures. They have to be separated out before we can use them. This is done by a form of distillation using the physical property of boiling points.

Fractional distillation

If you heat crude oil enough it will boil and turn to gas. In an oil refinery, crude oil is heated strongly in a furnace so that it all boils. The gas then passes up through a tower, which is hot at the bottom but cold at the top.

Long-chain hydrocarbons boil at high temperatures. They don't have to cool very much before they turn back to liquid. They condense at the bottom of the tower.

Short-chain hydrocarbons boil at lower temperatures. They have to cool a lot before they turn back to liquid. They travel up the tower before they condense.

▲ If only it were this easy…

Question

a Bitumen collects at the bottom of the tower. Does bitumen have long or short chains?

The tower has trays that collect the condensed liquid at different levels. Each tray collects a liquid that contains hydrocarbons with a similar carbon chain length. The trays at the bottom have the longest chains, the trays at the top have the shortest. The liquids that collect at each level are called **fractions**. The process is called **fractional distillation**.

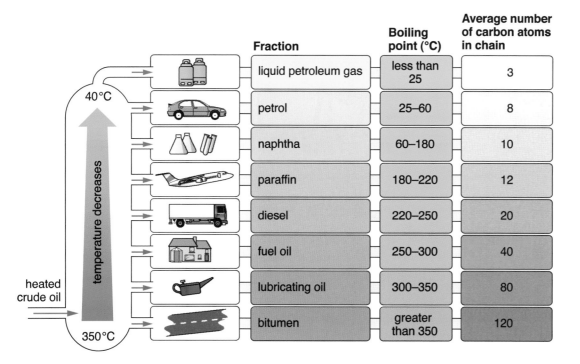

	Fraction	Boiling point (°C)	Average number of carbon atoms in chain
	liquid petroleum gas	less than 25	3
	petrol	25–60	8
	naphtha	60–180	10
	paraffin	180–220	12
	diesel	220–250	20
	fuel oil	250–300	40
	lubricating oil	300–350	80
	bitumen	greater than 350	120

heated crude oil

40°C

temperature decreases

350°C

Questions

b Sometimes fuel gets spilled onto the road. Which will evaporate faster, petrol or diesel?

c Will fuel oil have a higher or lower boiling point than lubricating oil? (Hint: how long are the carbon chains?)

The useful fractions

This process is not perfect as each fraction is still a mixture, but the hydrocarbons in each fraction have very similar properties. The hydrocarbons in petrol have carbon chains 6 to 10 atoms long.

The pie chart shows the amount of each fraction you get from crude oil.

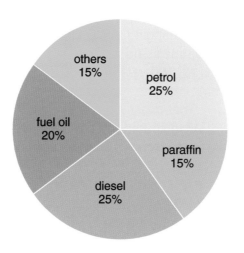

others 15%
petrol 25%
paraffin 15%
diesel 25%
fuel oil 20%

Question

d What percentage of crude oil has chains 40 carbon atoms or more long?

Key points

- The physical property of boiling point can be used to separate the mixtures in crude oil by fractional distillation.
- Long-chain hydrocarbons have high boiling points and condense better at the bottom of the tower. Short-chain hydrocarbons condense at the top of the tower.

Oil and global warming

Oil is mostly made from hydrocarbons. Hydrocarbons burn in air to form carbon dioxide and water. This reaction gives out lots of energy. We use this energy to keep us warm, power cars and generate electricity.

hydrocarbon + oxygen → carbon dioxide + water + **energy**

$$CH_4 + 2O_2 \rightarrow CO_2 + 2H_2O$$

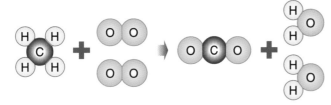

Question

a Which gas in air are hydrocarbons reacting with when they burn?

Carbon dioxide is the same gas that you breathe out – it can't be harmful, can it? Many scientists think that too much carbon dioxide could be causing **global warming**. Our climate does seem to be changing, but is carbon dioxide really to blame?

Scientists have shown that carbon dioxide traps heat in the atmosphere.

That's just as well. Without that effect the Earth would freeze!

But the amount of carbon dioxide is rising and the climate is changing. We've just had the hottest summers and most terrible storms for 100 years.

That doesn't prove the link. What caused the hot summers and storms 100 years ago?

Questions

b List two effects of global warming.
c What evidence is there that carbon dioxide is to blame?

Oil and acid rain

About 50 years ago, scientists in Europe started worrying about the environment. Trees were being damaged and fish were dying in lakes and rivers. Scientists found that the rain falling in these regions was a weak solution of sulfuric acid. They wondered what could be causing this **acid rain**. When they collected some data and showed it on a map, the source became clearer.

Power stations and factories burn oil or other fossil fuels. Oil often contains a little sulfur. Sulfur turns to sulfur dioxide when it burns. Sulfur dioxide is a poisonous, choking gas.

sulfur + oxygen → sulfur dioxide

Sulfur dioxide reacts with oxygen and water in the air to form acid rain. Today power stations remove sulfur dioxide from their waste gases. Petrol and diesel have to be 'sulfur-free' by law.

Question

d *Why does petrol have to be sulfur-free now?*

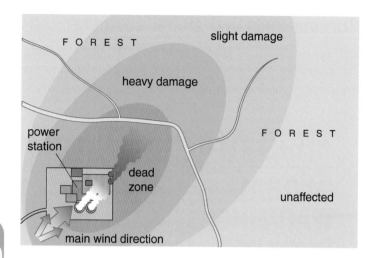

Question

e *Where is the source of pollution on this map?*

Oil and global dimming

When diesel burns it makes some soot – tiny particles of carbon. These can help to make summer smog in cities. They also get into the atmosphere and help clouds to form. The extra clouds reflect some of the sunlight back into space. As this pollution increases, the sunlight on the surface of the Earth gets weaker. This is called **global dimming**.

Hotter or colder?

Burning oil makes carbon dioxide, which warms the Earth up, but burning oil also makes carbon particles that cool the Earth down. Which effect is winning?

The Earth *is* getting warmer but without global dimming it could get much warmer. We want to stop the nasty pollution from diesel smoke but if we do the Earth could get hotter. Scientists are working hard to fully understand this problem so that we can solve it before it is too late.

Question

f *In Europe we now have laws to stop carbon particle pollution. What might this do to the average temperature in Europe?*

Key points

- Most fossil fuels contain a little sulfur. This combines with oxygen to produce sulfur dioxide, which causes acid rain.
- The carbon in fuels produces carbon dioxide, which causes global warming. Some carbon particles are also released from burning fuels, which causes global dimming.
- Sulfur can be removed from petrol and sulfur dioxide removed from waste gases to reduce pollution.

Burning fuels and global warming

Carbon dioxide is a greenhouse gas. We put 8 billion tonnes of it into the air every year when we burn fossil fuels. Many scientists think this is causing global warming but is it really that clear-cut?

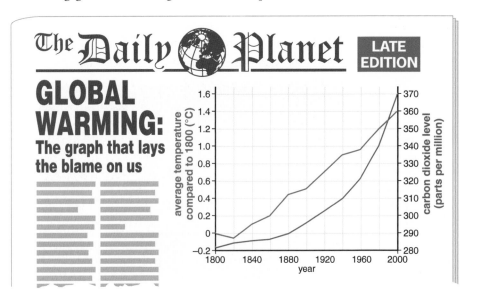

The Daily Planet LATE EDITION

GLOBAL WARMING: The graph that lays the blame on us

average temperature compared to 1800 (°C) / carbon dioxide level (parts per million) vs year (1800–2000)

Question

a How do the two lines on the graph compare? What do they show?

Natural carbon dioxide

There are many natural sources of carbon dioxide as well as what we produce.

- Natural wildfires give off carbon dioxide. Big fires in Indonesia in 1997 burnt for many weeks. They released 2 billion tonnes of carbon dioxide into the air.

- Erupting volcanoes release billions of tonnes of carbon dioxide into the air.

Wildfires and volcanoes do not occur to any pattern. It's hard to see how they fit in with the graph shown above.

Question

b After a wildfire, forests soon re-grow. What effect will that have on carbon dioxide levels over time?

Where does all this carbon dioxide go?

Scientists think that less than half of the carbon dioxide we make stays in the air. Let's look at where the rest goes.

- Plants absorb carbon dioxide during photosynthesis. A newly planted forest can take up to 1 tonne of carbon dioxide from the air per acre every year!

- Carbon dioxide dissolves in water. Much of the 'missing' gas has dissolved in the oceans.

Scientists are not exactly sure where all the rest of the carbon dioxide goes. Perhaps more of it is absorbed.

What can we do?

The link between burning oil and global warming is not completely proven, but many scientists think it is better to be safe than sorry. We should try to use less energy. This will reduce the amount of fossil fuels we burn. That may seem hard for us but is it fair to expect developing countries to reduce their use of fossil fuels? They are desperate to raise their standard of living by industrialising like we did in the 19th century.

Science to the rescue?

One possible answer is to trap the carbon dioxide before it gets into the air. One way to do this is to pump the gas back down old oil wells in the North Sea. Scientists are now trying experiments to see if this will work.

▶ Could old oil or gas wells save us from global warming?

> **Question**
>
> **c** Why is it a good idea to plant more trees?

> **Question**
>
> **d** List three ways in which you could save energy.

> **Key points**
>
> - Burning fossil fuels has contributed to an increase in carbon dioxide levels.
> - There are natural sources of carbon dioxide, such as wildfires and volcanoes.
> - Some carbon dioxide is absorbed by plants and oceans.
> - We still need to try to reduce our fossil fuel use to reduce the impact on the environment.

A thing of the past?

We've used up nearly half of our oil reserves. Pessimists see this as a 'glass half empty' and they fear what will happen when the oil runs out. Optimists see it as a 'glass half full' and that science will find a way to replace the oil! Where do you stand?

Type of oil reserve	Billions of barrels of oil
oil used in the last 100 years	620
known oil reserves	1050
estimated reserves as yet undiscovered	450

Questions

a List three problems of oil running out.
b Suggest two ways to make our oil reserves last longer.
c From the table it looks like we have enough oil to last another 200 years. Why is that not the case? (How much oil do we use now compared to 50 or 100 years ago?)

New fuel – hydrogen

Whatever happens, oil will run out eventually, so scientists are looking for new sources of energy. Hydrogen can be used as a fuel. When it burns it just makes water, so it is completely pollution-free.

hydrogen + oxygen → water + **energy**

Question

d Why do you not get carbon dioxide when hydrogen burns?

Some school bus fleets in California have already converted their buses to run on hydrogen.

▶ Pollution-free transport for the future!

However, there are problems with hydrogen. It has to be compressed and can be quite dangerous to store if not handled properly. Also, it can only be made in large amounts today by separating the hydrogen in water with electricity. Unfortunately the electricity is made by burning fossil fuels in a power station...

Ethanol

Many scientists are looking for renewable resources to replace oil. They have focused on plants as they trap energy from sunlight. Plants contain stored energy that we can use.

Sugar cane grows very quickly in hot countries such as Brazil. The sugar cane is fermented to produce **ethanol** – the alcohol you get in beer. This is then distilled and Brazilians run their cars on ethanol instead of petrol.

Many other plants store energy in their seeds as oil – olives, sunflowers, soya beans and so on. These oils can be made to burn well in engines. This new fuel is called **biodiesel**. You will find out more about it in Section 5 of this subject area.

Question

e Explain why hydrogen-powered buses are not as pollution-free as they seem.

Question

f List two ways in which we use the stored energy in plants now.

Question

g Ethanol is distilled, renewable fuel but large areas of forest in Brazil have been cleared to grow sugar cane. Draw up a list of advantages and disadvantages of using ethanol instead of petrol.

Key points

- Oil will run out one day but scientists have already developed new fuels, such as ethanol and biodiesel.
- The use of hydrogen as a fuel is still being developed.
- Some countries are already making good use of renewable fuels such as ethanol.

1   Match words **A**, **B**, **C** and **D** to the sentences in the table.
   A calcium hydroxide
   B carbon dioxide
   C calcium oxide
   D calcium carbonate

1	This is the chemical name for limestone.
2	This alkali is used to neutralise acid soil.
3	This solid forms when limestone is heated strongly.
4	This gas forms when limestone is heated strongly.

2

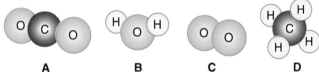

A   B   C   D

Match the diagrams **A**, **B**, **C** and **D** to the descriptions below.

1	This shows a molecule of oxygen, O_2
2	This shows a molecule of methane, CH_4
3	This shows a molecule of carbon dioxide, CO_2
4	This shows a molecule of water, H_2O

3   The formulae of four common compounds are
   1 quicklime CaO
   2 iron ore Fe_2O_3
   3 ammonia NH_3
   4 sulfuric acid H_2SO_4

   a Which compound contains calcium? *(1 mark)*
   b Which compound contains five atoms? *(1 mark)*
   c Which compound contains three different elements? *(1 mark)*
   d Which compound does not contain oxygen? *(1 mark)*

4

This building is made from concrete and glass.

   a Link the raw material to the product (you can use them more than once). *(2 marks)*

Raw material	Product
limestone	cement
clay	glass
sand	

   b What is cement mixed with to make concrete? *(1 mark)*
   c What problem might you face working in a glass-fronted building like this on a sunny day? *(1 mark)*
   d Suggest one advantage of building with concrete rather than blocks of limestone. *(1 mark)*

5   Limestone ($CaCO_3$) breaks down to quicklime (CaO) when heated. Zoe, Mumtaz and Tomi performed experiments to see how much quicklime they could get by heating 10 g of limestone. They each repeated their experiment three times. Here are their results.

| | Quicklime produced from 10 g of limestone (g) | | | |
	Expt 1	Expt 2	Expt 3	mean
Zoe	5.6	5.7	5.5	5.6
Mumtaz	5.62	5.66	5.64	5.64
Tomi	5.62	5.68	6.5	5.9

   a What other product is made in this reaction, alongside the quicklime? *(1 mark)*
   b Which student had been given an older, less precise balance   work with? *(1 mark)*
   c i Which student appears to have obtained the most reliable results? *(1 mark)*

ii Explain your answer to **i**. *(1 mark)*

d i Which student ran out of time and didn't heat her final piece of limestone long enough? *(1 mark)*

ii Explain your answer to **i**. *(1 mark)*

e The theoretical amount of quicklime produced is 5.6 g. Whose final answer appears to be the most accurate? *(1 mark)*

f A detailed analysis of the limestone used shows that the residue after heating is indeed slightly greater than 5.6 g. Suggest a possible reason for this. *(1 mark)*

6 Match words **A**, **B**, **C** and **D** to the numbers on the flow chart for tin production.

A reduced **B** crushed
C mined **D** separated

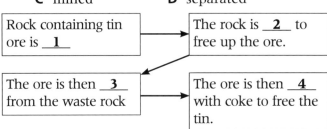

Rock containing tin ore is **1**

The rock is **2** to free up the ore.

The ore is then **3** from the waste rock

The ore is then **4** with coke to free the tin.

7 Choose the correct word or phrase from each pair to complete these four sentences,

a Iron, zinc and copper are (**alkali/transition**) metals. *(1 mark)*

b Copper is used for water pipes because it is (**easy to shape/conducts electricity**). *(1 mark)*

c Steel is an (**element/alloy**) containing iron and a little carbon. *(1 mark)*

d Brass is an alloy of zinc and (**iron/copper**). *(1 mark)*

8 Myanmar is a poor country with underdeveloped heavy industry. Bamboo grows well in its hot climate.

In London, scaffolding is built from steel tubes that are screw-clamped together. In Myanmar, scaffolding is built from bamboo poles lashed together with natural string.

a Which do you think would be stronger, the steel or bamboo? *(1 mark)*

b What would you notice if you picked up a steel pole and a bamboo pole? *(1 mark)*

c Suggest two reasons why bamboo is used in Myanmar rather than steel. *(2 marks)*

d Broken bamboo poles are simply thrown away. Why is that not a problem? *(1 mark)*

9 Look at the table:

Metal	Melting point (°C)	Strength	Cost	Density
aluminium	660	medium	medium	low
steel	1540	high	low	high
titanium	1670	high	high	low
tungsten	3400	high	medium	high

For each use below, suggest a suitable metal and give a reason (from the property table).

a the barrel of a Bunsen Burner *(1 mark)*

b the wing of a supersonic fighter *(1 mark)*

c the filament in a light bulb *(1 mark)*

d a commercial aeroplane *(1 mark)*

10 The permitted levels of some metal ions in drinking water are:

copper	1 mg per litre
lead	0.05 mg per litre
zinc	5 mg per litre

a From these figures, which metal is most toxic, which is least toxic? *(1 mark)*

Cattle in the fields around the river shown on the map became ill and metal poisoning was suspected. Water samples taken from the rivers at **A**, **B** and **C** were analysed and the results were:

	Concentration (mg per litre)		
	copper	lead	zinc
A	0.05	0.001	0.05
B	8.0	5.0	10.0
C	1.6	1.0	2.0

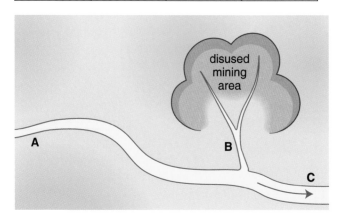

b Which site would give safe drinking water (in terms of metal content)? *(1 mark)*

c i Which site shows the most polluted water? *(1 mark)*

ii Where do you think this pollution has come from? *(1 mark)*

d i The rivers flow from **A** and **B** to **C**. Why are the metal levels lower at **C** than at **B**? *(1 mark)*

ii From the figures, which river carries more water, **A** or **B**? *(1 mark)*

e i One herd of cattle could only get to the river to drink at **C**. Which metals could be responsible for the poisoning? *(2 marks)*

ii Which metal is most likely to be responsible? Explain your answer. *(1 mark)*

11

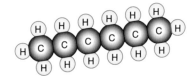

The picture shows a compound found in oil.

Match words **A**, **B**, **C** and **D** to the sentences in the table.

A carbon dioxide   **B** sulfur
C hydrocarbons   **D** carbon

1	Compounds like the one shown are called _____
2	When this compound burns in air it forms _____ and water.
3	This impurity found in oil leads to acid rain when the oil is burnt.
4	The soot that forms in car exhausts and may cause global dimming is made of _____.

12 Crude oil is split into fractions by distillation

Choose from the phrases below to label **1–4** on the diagram.

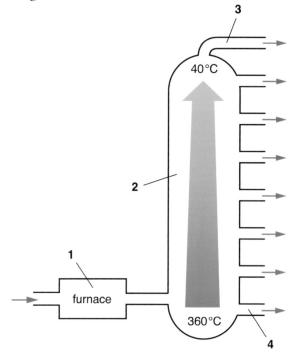

A short-chain hydrocarbons out
B crude oil vapourised
C lubricating oil
D the vapour rises and cools

13 Petrol and lubricating oil contain hydrocarbons. Petrol is runny and catches fire (ignites) easily. Lubricating oil is viscous and does not ignite easily.

Choose from the list the **two** statements that are true.

A Petrol is less volatile than lubricating oil.

B Petrol has shorter-chained hydrocarbons than lubricating oil.

C Petrol has smaller molecules than lubricating oil.

D Lubricating oil is a better fuel than petrol.

E Lubricating oil has a lower boiling point than petrol. *(2 marks)*

14 Most forms of transport work by burning fossil fuels. The table shows approximately how much carbon dioxide is produced for a 1 km journey.

by bus	1000 g
by large car	300 g
by small car	150 g
walking	15 g

a Why are we worried about how much carbon dioxide we produce? *(1 mark)*

b Why is it better for the environment to drive a small car? *(1 mark)*

c Some firms run 'car pools' where people take it in turn to give each other lifts. How does this help? *(1 mark)*

d A bus can take 50 people. How much carbon dioxide is produced per person per kilometer? *(1 mark)*

e Why isn't it zero for walking? *(1 mark)*

15 You need to know exactly how strong concrete is, to make sure that buildings do not collapse. Scientists test different mixtures to find out which is best for the job. One simple test squashes concrete samples in a vice until they break. Here are some results. The more 'turns' of the vice needed, the stronger the concrete.

Mixture	% cement	% sand and gravel	turns to break
A	20	80	3.4
B	15	85	2.4
C	10	90	1.7

a Which of these mixtures would you use to build a skyscraper? *(1 mark)*

b Cement is much more expensive than sand or gravel. Which mix would you use to make a garden path? *(1 mark)*

c From the table you might think that 100% cement is the strongest of all. This is not the case. How could you extend this experiment to find the strongest mixture? *(1 mark)*

d These tests were done a day after the samples were made. A sample of mixture **A** was tested a week later and took 5.6 turns to break it. Suggest a reason why new concrete bridges are often left for a week or so before traffic is allowed on them. *(1 mark)*

16 The sentences below are about how metals are produced. Choose the correct word or phrase from each pair given.

a Iron can be made from iron oxide in a blast furnace using coal because it is (**more/less**) reactive than carbon. *(1 mark)*

b Gold was one of the first metals to be discovered because it is so (**reactive/unreactive**). *(1 mark)*

c Aluminium has to be made by electrolysis because it is (**more/less**) reactive than carbon. *(1 mark)*

d Copper is *purified* by (**electrolysis/carbon reduction**). *(1 mark)*

17 Iron is produced in a blast furnace. The raw materials are coke, iron ore and limestone.

a When iron oxide is converted to metallic iron in this way the iron is
  A oxidised   B reduced
  C electrolysed   D neutralised

b A blast of hot air is used to
  A blow out the waste gases
  B oxidise the iron
  C help to form the slag
  D make the coke burn fiercely

c When the carbon dioxide reacts with more carbon it forms
  A sulfur dioxide   B metallic iron
  C coke   D carbon monoxide

18 a Choose words from the list for each of the spaces **1–4**.

water   carbon dioxide
  hydrocarbons   carbon

Coal is mostly made from __1__ atoms. When we burn coal the waste gas is __2__. Oil and gas are made of __3__. When we burn them we get carbon dioxide and __4__.

b Which sort of power station will be worse for the environment (in terms of global warming), coal-fired or oil-fired? Explain your answer. *(2 marks)*

c China is developing very fast and needs a lot of power for its industry. It has very large coal reserves to help it meet this demand. Why might that be a problem for the planet? *(2 marks)*

Oil is not just for energy!

You probably just think of oil as a good source of energy. In fact it has so many other uses that many people think it's a waste to just burn it!

Plastic fantastic

Where would we be without plastic? There would be no

- polythene shopping bags
- PVC 'artificial leather' sofas
- polypropene carpets
- children's bricks and toys
- polystyrene computer cases or packaging
- PET drinks bottles
- non-stick frying pans.

All these and many more products are made from crude oil. Oil is a truly versatile raw material.

Alcohol from oil

Industry needs alcohol. Ethanol is an alcohol used in industry. It is used as a solvent for varnishes, inks, paints and glues. All of this industrial ethanol is made from oil. We use a lot of these substances.

Smart plastics

There are more advantages. New plastics are being made with 'smart' properties.

- Some plastics can grow or shrink. They could be used as the muscles of future robots.
- Plastics used in electronics could soon give us giant TV screens that roll up like blinds.

Problems with plastics?

Plastics also have disadvantages. Plastics are easy to use in large-scale, automated industrial processes. This may be good for us as consumers because we get cheap goods but it has put many skilled craftsmen from carpenters to leatherworkers out of work. These people have to be retrained and find new jobs.

▲ A smart polymer future?

Because of plastics we now use fewer local renewable materials, such as wood and leather. Instead we rely more and more on expensive imported oil. Oil prices change rapidly and this can lead to economic instability. And what will we do when *all* of the oil finally runs out?

Plastics can also cause environmental problems. They are difficult and expensive to dispose of. You will consider this issue further in Section 4.5.

Think about what you will find out in this section

What are plastics and why do they have such useful properties?

How are plastics made from crude oil?

How is alcohol made from oil?

How can science help to develop fantastic new smart materials?

How do plastics affect the environment?

Crude oil fractions

The fractional distillation of crude oil makes many useful products but some of these products are more in demand than others.

Petrol is used more than any other fraction. You *could* make enough petrol by distillation alone, but this would make too much long-chain oil.

Question

a *Why do you think petrol and diesel are used more than other fractions?*

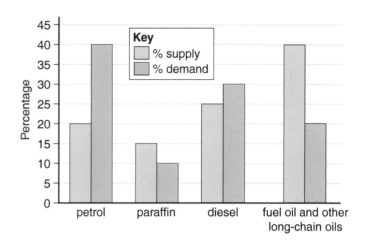

Get out the scissors?

Fortunately, it is fairly easy to 'cut up' long-chain hydrocarbons. If they are heated, they can be broken up. This thermal decomposition reaction is called **cracking**.

Industrial cracking

The cracking process happens at the oil refinery.

- The unwanted long-chain fractions are boiled.
- The vapour is passed over a hot catalyst. (This makes the reaction work faster.)
- The long-chain molecules break up into smaller pieces.
- The shorter-chain products are then sent back up the fractionating tower to separate them.

The extra petrol needed is produced in this way, as well as other important chemicals.

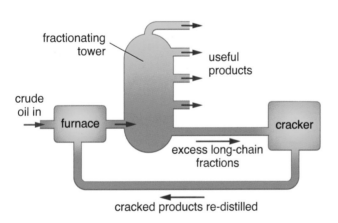

Question

b *Why do you think the products from cracking need separating again?*

More about cracking and its products

You have to break one of the chemical bonds between two carbon atoms to crack an alkane. The pieces of broken alkane are very reactive. They want to make new bonds. The bigger piece bonds with a hydrogen atom from the smaller piece to form a shorter alkane. The little bit left now has two broken bonds, as one of its hydrogen atoms has been pulled off. These snap together to make a double bond between two carbon atoms.

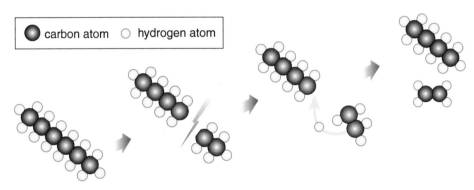

▲ Cracking alkanes.

Hydrocarbons with double bonds like this are called **alkenes**. They have two hydrogen atoms less than the saturated hydrocarbons we saw in Section 3.1. Because of this they are called **unsaturated** hydrocarbons. The simplest alkene, with just two carbon atoms, is called ethene.

At first, ethene was simply burnt off to get rid of it. Now scientists have discovered lots of fantastic uses for ethene!

Ethene's chemical formula is C_2H_4. We can also draw alkenes to show the chemical bonds. The carbon-to-carbon double bond is shown as C=C.

▲ Some simple alkenes.

Questions

c Write down how many carbon and hydrogen atoms there are in each of the alkenes shown above. Then write down the formula for each.

d What is the pattern in the number of carbon and hydrogen atoms in alkenes?

e What is the chemical formula of an alkene with 12 carbon atoms?

Key points

- Hydrocarbons can be cracked by thermal decomposition using a catalyst to produce smaller, useful molecules.
- The products of cracking are unsaturated alkenes. Ethene, C_2H_4, is an example. Double the number of carbon atoms to get the number of hydrogen atoms. The general formula is C_nH_{2n}.
- One of the fractions from cracking is petrol, which is much in demand.

Traditional alcohol

Ethanol is the alcohol in wines, beers and spirits. These are made by fermenting grape juice or malt mash with yeast. This makes a weak solution of ethanol (alcohol) in water. This may be distilled to concentrate the alcohol to give a stronger alcoholic drink such as whisky.

Question

a *What effect does ethanol have on people?*

Ethanol in industry

▼ Methylated spirits are used as a solvent to clean up paint or glue spills at home.

85% ethanol   40% ethanol

Question

b *'Meths' has chemicals added to the ethanol to turn it purple and give it a nasty, bitter taste. Why do you think this is done?*

Fermented alcohol is expensive because it is made in small amounts and the government puts a lot of tax on it. Industrial ethanol is pure (100%) and is a very important industrial chemical. It is used as a solvent in glues and paints, for example. Industry needed a cheaper source of ethanol and oil provided the answer.

Ethene is a waste product when oil is cracked. Scientists soon discovered a wonderful use for this. They made it react with steam to give ethanol.

     ethene + water → ethanol

Question

c *Why does industry prefer this method of making ethanol?*

So which method is best?

It depends what you want the ethanol for – and whether you are thinking about now or the future.

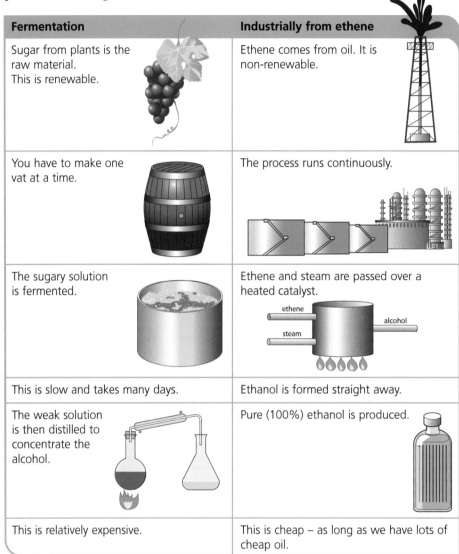

Fermentation	Industrially from ethene
Sugar from plants is the raw material. This is renewable.	Ethene comes from oil. It is non-renewable.
You have to make one vat at a time.	The process runs continuously.
The sugary solution is fermented.	Ethene and steam are passed over a heated catalyst.
This is slow and takes many days.	Ethanol is formed straight away.
The weak solution is then distilled to concentrate the alcohol.	Pure (100%) ethanol is produced.
This is relatively expensive.	This is cheap – as long as we have lots of cheap oil.

Industrial ethanol is made from ethane today because it is so cheap, but the price of oil will rise as it runs out. Eventually there will be none left. Plants are a renewable resource. Fermentation can continue for as long as the sun shines.

Question

d What is the big advantage of:
  (i) the fermentation method (ii) the industrial method?

Key points

- Ethene can be reacted with steam using a catalyst to produce an industrial alcohol, ethanol.
- The industrial method relies on oil and the by-product of cracking, ethene, which comes from non-renewable oil.
- The fermentation method relies on renewable plant resources.

Polymerisation

Alkenes such as ethene made by cracking have many other exciting uses. These small molecules can be made to 'pop' together like beads to make very large molecules. There may be thousands of these small molecules, called **monomers**, popped together in a long chain, called **polymers**. 'Poly' means many. These long chains stick together like a tangled mass of spaghetti. This process is called **polymerisation**.

Polymers are also called plastics. They are very easy to shape to make cases for radios or sink units. Plastics are very versatile and have many other uses.

> **Question**
>
> **a** List three uses of plastics in the photograph of the modern klitchen below.

▼ How plastics have changed our lives.

Poly(ethene)

Polymers are named by putting 'poly' in front of the name of the small molecule. The simplest molecule is called ethene, so the polymer made from this is called poly(ethene). This is often shortened to polythene.

> **Question**
>
> **b** Another small molecule is called styrene. What is the polymer made from this called?

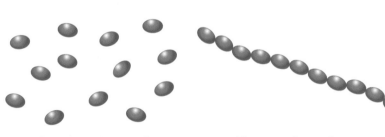

Cracking makes small ethene molecules.

These can be made to pop together to form poly(ethene).

The chains stack up like molecular spaghetti.

Poly(ethene) is very easy to shape. It can be moulded into bottles for drinks. It can also be rolled into thin, flexible sheets. This is ideal as a food wrap or for plastic bags.

Question

c What advantages do poly(ethene) bottles have over glass ones?

Poly(propene)

Poly(propene) is a tougher plastic than poly(ethene). It's not so flexible. It is used to make hardwearing things such as bowls, crates or even school chairs. It is also made into fibres for carpets and ropes. It is safe and can be made brightly coloured so it is often used for children's toys.

Question

d What alkene is poly(propene) made from?

The polymer for the job

Different polymers have different properties that make them suitable for different jobs.

Polymer	Flexibility	Toughness	Relative cost	Other property
poly(ethene)	high	low	low	
poly(propene)	medium	high	medium	easy to colour
poly(styrene)	low (brittle)	brittle	medium	can be precisely moulded
poly(chloroethene) (PVC)	high	high	high	resistant to corrosion a very good electrical insulator

Question

e Which polymer would you use for these products and why:
(i) insulation on an electrical cable
(ii) the case for a stereo system
(iii) a carrier bag
(iv) a washing-up bowl?

Smart plastics

Scientists are continually developing new plastics with strange new properties. Shape memory polymers stretch like rubber when warm but will stay in their new shape when cooled. Warm them up again – and they snap back to their original shape.

Question

f Shape memory polymers can be used to make moulds for casting plaster or concrete. You can make the mould by warming the polymer, stretching it into shape and then cooling it. Explain how this material can be used over and over again for different moulds.

Key points

- Alkene monomers such as ethene and propene can be used to make polymers such as poly(ethene) and poly(propene) in a process called polymerisation.
- Polymers have many uses, depending on their properties.
- Scientists are now developing 'smart polymers' with strange but useful properties.

From the slime monster to hydrogels

Slime is wonderful stuff. But what is it? And how can you change it from soft dripping slime to bouncing semi-rubber?

Simple slime is made from PVA glue. PVA is short for poly(vinyl alcohol) – the old name for poly(ethenol). On its own it is not strong, but it mixes with water to make glue.

If you mix PVA glue with borax, chemical reactions make 'cross-link' bonds *between* the polymer chains. This makes a loose grid. Water molecules can get trapped in this grid and are held in place. This gives slime its properties. Slime with few cross-link bonds is very runny. Slime with more cross-link bonds is much more viscous.

Its open structure and trapped water makes slime soft and easy to pull apart by breaking the cross-link bonds between the chains. When two pieces of slime are pushed back together, the cross-link bonds reform and the slime forms a single piece again.

Materials like slime are called **gels**. You can make weak gels like slime by including lots of water. If you have more polymer and more cross-link bonds you get bouncier and tougher gel.

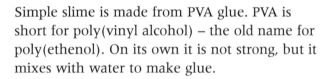

PVA: polymer chains –
no cross-links

PVA glue: polymer chains –
no cross-links, water molecules
held loosely between chains

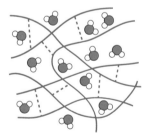

slime: polymer chains –
weak cross-links, water
molecules trapped loosely
between chains

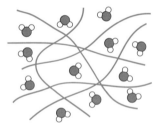

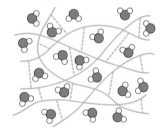

hydrogel: polymer chains –
strong cross-links, water
molecules trapped
between chains

Question

a Jelly has a similar structure to slime. What's different about its properties?

Some polymers make gels with much stronger cross-links. These **hydrogels** are used to make soft contact lenses. Different hydrogels can be used as dressings in hospitals to cover wounds. They keep the wounds safe from infection. Hydrogels keep wounds clean but let in oxygen and water, which means the wound heals more easily.

> **Question**
>
> **b** Compare a normal plaster with hydrogel dressings. What are the disadvantages of normal plasters?

Keeping the water out

The polymer PTFE (polytetrafluoroethene) is used to make non-stick frying pans. It is also used to make clothing waterproof. A layer of PTFE has millions of tiny pores. When you sweat, the water molecules escape through these pores. The layer 'breathes' like cotton. But when it rains the water droplets are much too big to get through the pores.

> **Question**
>
> **c** Polythene is waterproof but has no pores. Why do we not just wear polythene clothes to keep dry?

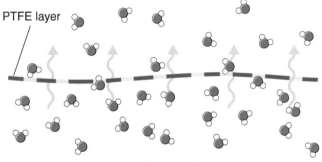

water vapour from sweat can escape

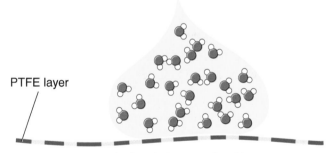

water in raindrops can't get in

Futuristic applications

Scientists have now made plastics that conduct electricity and can be used in microchips. Plastic microchips are easy to make. You can print them using special inks on special plastic sheets. All you need is a big bubblejet printer. Some other plastics glow in different colours. They can be used for decoration or emergency lighting.

Soon smart polymers will be everywhere. Athletes will have computers built into their clothing to see how they are performing. You may have computers built into your clothing that check your health!

> **Key points**
>
> - The properties of polymers depend on what they are made from and how they are produced.
> - Polymers share many useful applications and new uses are being developed all the time. Hydrogels are useful new polymers.

Aren't plastics great?

Plastics are everywhere today. Much of our food now comes pre-packed in plastic. We have plastic bottles for our drinks, polystyrene cartons for our hamburgers, plastic trays and clingfilm for meat, and plastic foam packaging.

The downside

We use plastic for many things these days, but how do we get rid of it all? Over a million tonnes of plastic packaging are produced every year in Britain. A lot of this ends up as litter on our beaches and in the countryside, an ugly reminder of our wastefulness.

We throw paper away too, but this natural material rots down rapidly in the soil. It is **biodegradable**. Plastics, however, are not biodegradable, and may take tens or even hundreds of years to break down. Most plastic rubbish ends up in the dustbin. It then goes to the local tip, but these waste tips are filling up fast.

Different people, different solutions

We are being buried alive by plastic that pollutes our environment. Plastic bags and bottles are strewn along our roadside. We must ban all plastic packaging now before it's too late. What's wrong with paper bags? They use them for shopping in America. People should take shopping bags with them to supermarkets like they do in Holland and Germany.

Earnest Green, environmental activist

Plastics are made from non-renewable oil. What a waste to just throw it all away. We must recycle wherever possible. I got my local council to set up plastic recycling bins. Not all manufacturers mark their containers with the plastic type. I'm campaigning to try to make this compulsory.

Jayne Middleton, local campaigner

Smarter technology can solve the problem. There are jobs for scientists working on the development of new plastics that are biodegradable. They would quickly rot away without leaving toxic waste. They would be more expensive than oil-based plastics but it would be worth it to stop damage to the environment. And the price would soon come down if we made more of them.

Susan Foresight, chemist

Recycling uses a lot of energy – often more than it would take to make new plastic – so it's a lot of effort for little return. But plastic waste contains as much stored energy as fossil fuels. So why bother to recycle when you can just burn the plastic waste as fuel for a power station? That way you clean up the environment and save oil at the same time.

Jack Burnham, combustion engineer

Question

d *What do you think?*
Draw up a table like the one below and use what people say on this spread to help you list the social, economic and environmental impacts of plastics and natural materials.
Design a poster to make people think about this problem, using the points you have noted.

	Social impact	Economic impact	Environmental impact
Plastics			
Natural			

Key points

- Many polymers are not biodegradable, which causes a problem with waste disposal.
- The disposal or recycling of polymers costs money and can have social, environmental and economic impacts.

Plants make glucose by photosynthesis, but some change this glucose to oil. Oil is a more concentrated store of chemical energy than glucose. Oil in seeds provides energy for growing seedlings. We can eat these seeds and use the energy directly, or we can squeeze the oil out of them and burn them as fuels.

Extracting the oil

Olives have been used to make oil for thousands of years all around the Mediterranean. Traditionally, the olives were crushed by giant stone wheels rolling over them, or squashed in large, hand-turned presses. Now these are being replaced by industrial hydraulic presses. The oil is squeezed out of the crushed pulp, runs out and is collected. The oil is then separated out from any water or other impurities. Even with modern methods olive oil is still expensive as olives are difficult to harvest and contain less than 20% oil. Today, sunflowers are also widely grown for oil.

Oil is sometimes removed from plants by distillation, but this can alter the flavour and smell of edible oils. Some plants, like lavender, produce small amounts of scented oil. This is often removed by distillation with water, which works at a lower temperature.

Oilseed crops

In Southern Europe, sweetcorn and sunflowers grow well. In cooler Britain, the favourite oilseed crop is rapeseed as it grows well in our climate and the seeds contain up to 50% oil by mass. This close relative of cabbage with its sickly yellow flowers has changed the colour of the British countryside in summer over the last 20 years.

Rapeseed oil is produced cheaply by industrial-scale processes. It is also a 'healthy' oil in your diet (see Section 5.1). The seeds are scattered easily and you can now see this plant growing along roadsides and invading fields. There is a danger that it will push out native wild flowers.

Vegetable oil for transport?

Oils are fuels for our bodies. Vegetable oils can also be made into **biodiesel** as a fuel for cars and lorries. Biodiesel gives off carbon dioxide when it burns, but the plants that it is made from took in carbon dioxide when they grew so it is a renewable fuel that does not cause global warming – a possible solution to our future fuel needs.

However, you would need to grow a lot of plants. You get about 1000 litres of oil from $4000\,m^2$ of oilseed rape. That's enough to run an average car for a year. But there are 24 million cars in Britain. You would need to turn $96\,000\,000\,000\,m^2$ of land over to oilseed farming if you wanted to replace petrol and diesel. In Britain that would not leave enough land to grow our food!

▼ Filling up with biodiesel.

Think about what you will find out in this section

Which plants provide vegetable oil and how do their oils differ?

How are oils changed physically or chemically before use?

Could vegetable oils become the fuels of the future?

How can we evaluate our increasing use of vegetable oils for fuels and food?

What additives are used in processed foods?

How can scientists check what's in processed foods?

How can science help us to understand more about healthy eating?

Healthy or not?

Confusing, isn't it! Are oils good for us or not? Science can help you.

▲ Oily food is bad for you!

▲ Oily food is good for you!

Oils for energy

Oils and fats are energy food. Fats are just solid oil. Every gram of oil provides 39 kilojoules of energy – twice that from carbohydrates. On just 10 g of oil you could:

● sleep for 1 hour ● walk for 30 minutes ● cycle for 15 minutes ● run for 8 minutes

You must have some oil in your diet for your body to work properly. For example, you get vitamin A from oily foods. Foods fried in oil are crisp and tasty, but fried food also absorbs a lot of oil. People who eat too much fried food can get overweight. There are other health problems that depend on the type of oil you eat.

What sort of oil?

Vegetable oils have long carbon-chain molecules. In some oils, all the C–C bonds are single bonds. These are saturated oils. Animal fats are also saturated like this. Eating too many saturated oils or fats can build up a fatty chemical called cholesterol in the body. This can block arteries and cause heart disease.

Vegetable oils also contain some molecules with C=C double bonds – unsaturated oils. These seem to be much better for our health. Olive oil and sunflower oil have a lot of unsaturated oil.

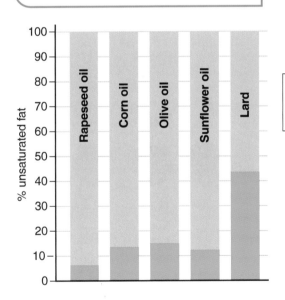

▲ Butter contains saturated fats, but spreads made from olive oil or sunflower oil are unsaturated.

Key

saturated   unsaturated

How can you tell which is which?

The C=C double bonds in unsaturated oils make them more reactive than saturated oils. They react with orange–brown bromine water, which loses its colour. A similar reaction occurs with iodine. These reactions are used as a test for unsaturated oils.

Changing vegetable fat

Fresh food may be best for our health, but processed foods are certainly quicker. What happens to food when it is processed?

Most natural fats come from animals. They are better than oils for making cakes or pastry because they are solid, but they are linked to heart disease.

Vegetable oils are mostly unsaturated. Animal fats are mostly saturated. Scientists have found a way to change cheap unsaturated vegetable oils into saturated fats for cooking.

Unsaturated oil can react with hydrogen if the oil is warmed to 60 °C and hydrogen gas is bubbled through it. The double bond snaps open and bonds with a passing hydrogen molecule to become saturated. Nickel is used as a catalyst to speed up the reaction. This process is called **hydrogenation**. Hydrogenated oils are used in margarine, sometimes used to replace butter because it is spreadable at room temperature. Hydrogenated oils are also used in cakes and pastries – and in chocolate!

▲ ▼ Slave over a hot stove… or pop it in the microwave?

Questions

a Why are most fats not suitable for vegetarians?

b Do you think that hydrogenated oil is as good for our health as the original unsaturated oil? Explain your answer.

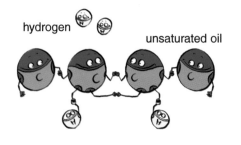

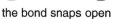
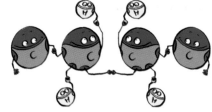

hydrogen　　unsaturated oil

the bond snaps open　　and grabs the hydrogen to become saturated

What is added to our food?

Processed foods can lose some of their flavour and natural colour. They also need to last a long time because they have to travel to the supermarket and be displayed there for a while without going off. Processed foods have additives put in to overcome these problems.

Question

c Which group of additives do you think is the most important?

● **Flavour:** Sugar and salt in our baked beans improves the flavour. Monosodium glutamate is another additive that makes food tastier. It is often used in Chinese food.

Colour: We like our peas green and our tomatoes red, but processing dulls these colours. Some products use food colours to make the food look nicer.

Preservatives: Vinegar for pickle and sugar for jam have been used for centuries. Chemicals such as benzoic acid and sulfur dioxide are newer alternatives. Preservatives stop microbes growing in the food.

Are all these additives good for us?

Used wisely, additives can help us make the most of our food. They can stop food going bad and make our food look and taste nicer.

There are also some problems with additives:

- Some people might be allergic to some additives.
- They might be used to disguise poor-quality ingredients.
- Some bright colours used in sweets and drinks may cause children to lose concentration and become hyperactive.

How can we tell what's in our food?

By law all processed food products have to list their ingredients. Some have been given E-numbers by the Government to show that they are allowed. For example, E100 and E199 are colours and E200 and E299 are preservatives. Some people don't like having any chemicals in their food, but many additives are perfectly natural. E300 is vitamin C whilst E260 is acetic acid – vinegar.

TEARAWAY TOTS TANKED UP ON TARTRAZINE

My little Johnny became a monster

Questions

d Do you think the benefits of additives outweigh the drawbacks?

e This fizzy drink label shows three E-numbers. What are these for?

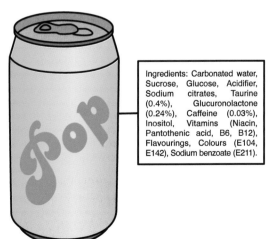

Ingredients: Carbonated water, Sucrose, Glucose, Acidifier, Sodium citrates, Taurine (0.4%), Glucuronolactone (0.24%), Caffeine (0.03%), Inositol, Vitamins (Niacin, Pantothenic acid, B6, B12), Flavourings, Colours (E104, E142), Sodium benzoate (E211).

Key points

- Vegetable oils can be hardened by hydrogenation so they can be used in spreads, cakes and pastries.
- Processed foods contain additives to improve taste and appearance, and to preserve them, but there are also disadvantages of using additives.
- Permitted additives are classified using the E-number system.

So what is an emulsion?

Many people drink a glass of white emulsion every day. Milk is a common emulsion.

Oil and water do not mix – the oil does not dissolve in the water. If you shake water and oil together tiny droplets of oil form in the water. An **emulsion** is a mixture of tiny droplets of oil in water, or water in oil. But the droplets soon separate out. This happens with a simple salad dressing made from oil and vinegar (a watery solution).

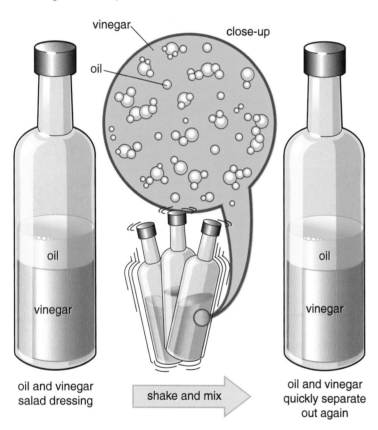

oil and vinegar salad dressing

shake and mix

oil and vinegar quickly separate out again

Emulsions don't always separate that easily. Milk is an emulsion of tiny droplets of fat in water. Cream is similar but has more fat and less water. In butter, the emulsion is the other way around – it has a few water droplets trapped in the fat.

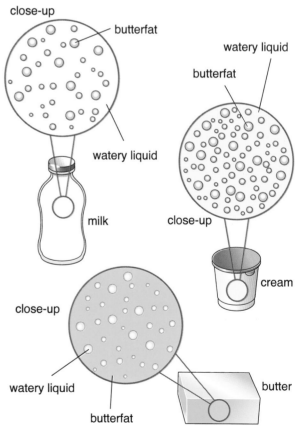

Question

a *Skimmed milk is milk that has had the cream removed. Does it have more or less fat than ordinary milk?*

Keeping them together

Oil and vinegar separate out quickly. Mayonnaise is made from oil and vinegar but it also contains egg yolk. The egg yolk stops the emulsion from separating out. Substances that do this are called **emulsifiers**. There are many different chemicals that can be used. They are given E-numbers too, from E322 to E499.

> **Question**
>
> **b** A jar of mustard mayonnaise contains E102 and E375. What are these for?

Why are emulsions useful?

Emulsions are thicker than oil or water. The thickness depends on the amount of oil and water, and how it is arranged. Their uses depend on their special properties.

Emulsion paints are easy to use. Many emulsion paints are non-drip because they are quite thick. The paint sticks to the brush and doesn't drip everywhere.

Food emulsions are also less runny. Mayonnaise is much thicker than salad dressing. It doesn't pour, it dollops – and sticks nicely to your chips. Cream is less runny than milk; double cream is even thicker as it has more fat. You can make it even thicker by whisking in some air bubbles. Whipped cream is very stiff. It sticks to your strawberries or can be piped into pretty shapes on your trifle. Ice cream is another water and fat emulsion with added air bubbles.

Emulsions are very important in the food industry. They are used to change the texture of the food:

- to help it coat other foods
- to make it keep its shape
- to give it a creamy, even texture.

Soft margarines are also water-in-oil emulsions. There is enough water in the margarine to make it soft enough to be spread straight from the fridge.

> **Question**
>
> **c** What property do non-drip paint and whipped cream have in common?

> **Question**
>
> **d** List three emulsions you have eaten this week.

> **Key points**
>
> - Emulsions are mixtures of oil and water, which can be made to stay together using emulsifiers.
> - Emulsions have different uses depending on their properties. Examples are milk, butter, mayonnaise and paint.

Food colouring

Fresh chillies can be bright red. Dried chillies are much duller. So why do these dried spices look so bright? They may have had food colours added to make them look fresh.

This is fine if the colours are harmless. But sometimes dyes are used that should not be put in food. Some dyes cause cancer. Should we worry about what's in our food?

> **Question**
>
> **a** List three foods that are coloured with food dyes.

SUDAN 1 FOUND IN PROCESSED FOOD

400 products taken off shelves due to cancer-causing dye

How can we tell what's in our food?

Foods are mixtures so to tell what's in them you have to first separate the mixtures.

One easy way to separate food colours is to use **chromatography**:

- Put a spot of food colour onto a piece of filter paper.
- Let water soak up through it.
- Wait until the colours have moved up the paper with the water and separated out.
- Match the colours to known dyes.

> **Question**
>
> **b** The yellow dye tartrazine (E102) makes Johnny very active and over-excited. Which sweets (A to E) should he avoid?

Y A B C D E   Y   Tartrazine

Reliability

Withdrawing food from supermarkets is very expensive. Supermarkets use laboratories that do chemical analyses to test foods. These laboratories have lots of complex machinery to do chromatography and other types of analysis. If you are going to test food and then withdraw it, you need to be sure you have got reliable results. You need to repeat your results and then take an average to reduce any errors. The machines must be precise enough to give very similar readings for the same test, time after time. They will also need to be calibrated carefully – tested on known standard samples – and adjusted so that the average result is accurate. You have to be sure that your results are repeatable – would you get the same result if the tests were repeated later or in a different laboratory?

Question

c *A standard 10g mass was weighed repeatedly on two balances, A and B. The results were A 10.1g, 10.0g, 9.9g, and B 10.01g, 10.02g and 10.03g. Which balance is the most accurate?*

Fortunately modern laboratories are very reliable. Special machines can make a very detailed analysis of a dye and match it to a database of dyes. Each dye has its own unique chemical fingerprint.

Question

d *Which of these three chilli powders, firebrand, demon or karakatoa, contains Sudan 1?*

Analysis in the lab

Modern laboratories are very high-tech and rely on automated systems. They are:

- fast: samples can be tested one after the other, 24 hours a day
- accurate: once calibrated these machines will give repeatable and reliable results
- sensitive: these systems can deal with microscopic amounts.

Question

e *The databases have 'fingerprints' for chemicals such as sulfur dioxide and carbon monoxide. Which group of scientists might want to test for these chemicals?*

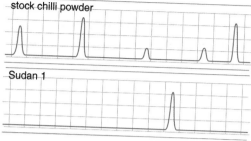

From database for comparison

stock chilli powder

Sudan 1

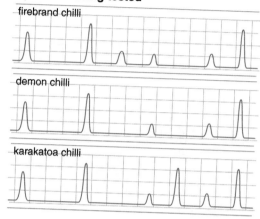

Chilli powders being tested

firebrand chilli

demon chilli

karakatoa chilli

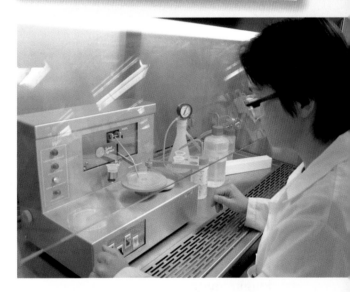

Key points

- Chromatography and other chemical analyses can help us find out what additives are in our food.
- Measuring instruments need to be calibrated so that they give accurate readings. Repeated readings are needed to give reliable measurements when averaged out.

Ever since people have had accurate maps of the world they have wondered at how some of the continents seem to fit together like a jigsaw puzzle. Perhaps you have wondered too?

People also wondered why the same types of land animals are found on different continents. Why are marsupials such as possums found in South America and Australia yet nowhere else? How could they have crossed deep oceans to get from one continent to another? Some scientists thought there must once have been land bridges that linked the continents.

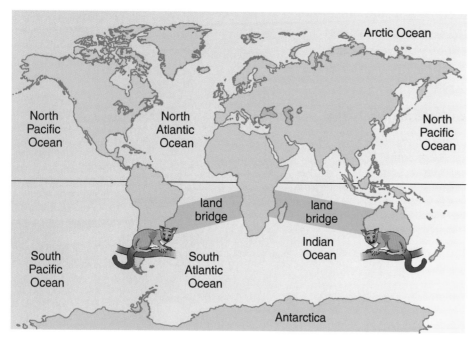

In 1912, a scientist called Alfred Wegener looked at the shapes of South America and Africa. The shapes fitted together like the two halves of a torn picture, but so did all of the details.

- The different rock types matched up.
- The same fossils were found on both sides.
- There was evidence of an ancient shoreline that matched.

He thought South America and Africa must have been part of a bigger continent that had split and moved apart. He called this process **continental drift**, but he didn't know how it worked. When his work was published in English translation in 1924 it immediately caused uproar. Most scientists thought the idea was nonsense.

- It went against the current ideas about land bridges.
- He could not explain why the continents moved.
- Physicists said (rightly) that it would be impossible for continents to 'plough through the crust' without breaking up.

Key
same fossils found today
shoreline 450 my ago
ancient rocks

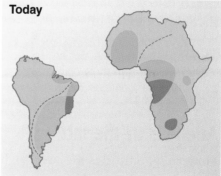

Today

▲ How the continents are today.

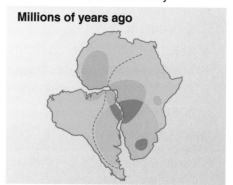

Millions of years ago

▲ How Wegener thought the land must have been millions of years ago.

The scientific community shut Wegener out. He couldn't even get a job as a professor in Germany. He had his great idea too soon, without enough evidence to back it up. He died in 1930, frozen to death on an expedition over the Greenland ice cap.

Unexpected mountains under the sea

One of Wegener's problems in getting his ideas accepted 80 years ago was that nobody knew what was at the bottom of the oceans. Then scientists started to survey the floor of the Atlantic Ocean, which is up to 7 km deep, for the first time. They were amazed to discover a chain of undersea volcanoes down the middle of the ocean! No one expected there to be mountains under the sea.

In the 1960s scientists realised the line of volcanoes was along a crack in the Earth's crust. Magma from the mantle pushed into this crack, cooled and set to rock. The sea floor was spreading apart and the ocean getting wider. This was how the continents moved apart.

The seal of approval

In 1964 the Royal Society in London had enough evidence to support Wegener's idea of continental drift. Wegener's explanations were not accurate, as the whole crust moved, not just the continents, but his idea finally gave birth to the theory of **plate tectonics** – the movement of plates or parts of the crust carrying continents.

▼ Iceland sits astride the mid-Atlantic rift. The central valley gets wider every time its volcanoes erupt.

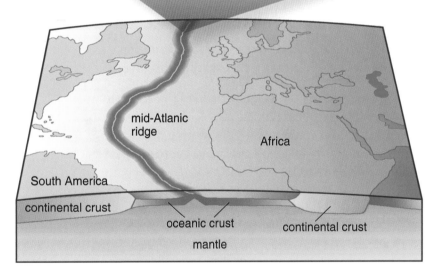

mid-Atlantic ridge

Africa

South America

continental crust

oceanic crust

continental crust

mantle

Think about what you will find out in this section

Why is the Earth so 'restless'?

How have the Earth and its atmosphere changed over time?

Why can it take a long time for some scientists' ideas to become widely accepted?

Can we use our understanding of the Earth's structure and processes to help save lives?

Can we use science to reduce the harmful effects of human activity on the atmosphere?

Beneath your feet

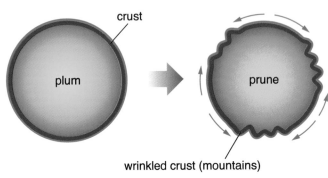

Peaceful Earth

You may think that the Earth is a peaceful place geologically, but don't be fooled. As well as the continents moving around, new mountains are pushed up and old ones are worn away, volcanoes are erupting in some areas of the world and earthquakes shake the ground.

> **Question**
>
> **a** Where have you heard of earthquakes occurring or volcanoes erupting recently?

Mountain building

Mountain ranges such as the Himalayas are built from folded rocks. A hundred years ago, some scientists thought that the Earth must have shrunk as it cooled down. They thought that this caused the crust to get wrinkles, just like a plum turning into a prune. They thought mountains were just big wrinkles! To understand how mountains are made, we need to know more about the inside of the Earth.

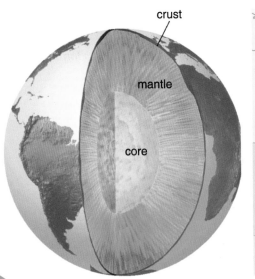

crust

plum → prune

wrinkled crust (mountains)

What is the Earth like below the surface?

We live on the surface of the Earth. This surface layer is called the **crust** – it is a thin hard layer.

Beneath the crust is a hot layer called the **mantle**. Over millions of years it can move. The continents on the crust and the top part of the mantle are divided into a number of large pieces called **tectonic plates**. These move, carrying the continents.

> **Question**
>
> **b** Iceland is a volcanic island in the middle of the Atlantic. Why is it getting wider?

▼ This map shows the main plates and how they are moving.

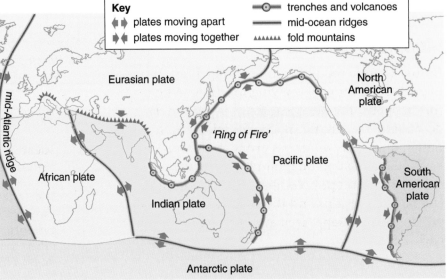

crust
mantle
core

Key
◀▶ plates moving apart
▶◀ plates moving together
⊸⊙⊸ trenches and volcanoes
—— mid-ocean ridges
▲▲▲▲▲ fold mountains

mid-Atlantic ridge
Eurasian plate
North American plate
'Ring of Fire'
Pacific plate
African plate
South American plate
Indian plate
Antarctic plate

Why do continents move?

If you heat a liquid from below, warm currents rise up and make the liquid swirl and mix. These are called **convection currents**.

The rocks in the mantle are being heated by natural radioactivity. This heating causes convection currents in the mantle. They are very slow currents but they make the tectonic plates move. They move very slowly, just a few centimetres a year, but they have been moving for hundreds of millions of years. Over such a long time, these tiny movements can make continents move around the Earth and force new mountains high into the air.

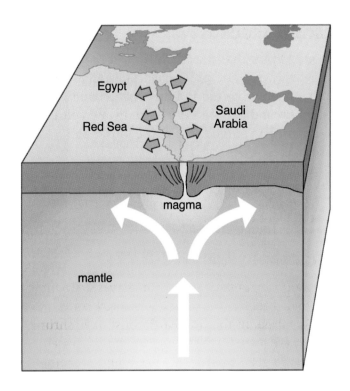

Question

c The Red Sea is on a plate boundary. The plates are moving apart. What will happen to the Red Sea over the next few million years?

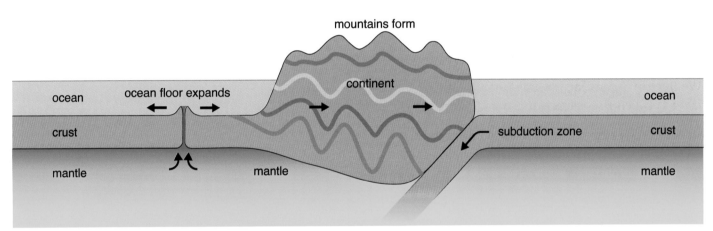

Mountain building cycle

In the Atlantic the plates are moving apart, but in the Pacific they are moving together. When this happens, the old ocean crust is pushed back down into the mantle. Rocks are folded up to form new mountains, volcanoes erupt and powerful earthquakes shake the ground. Eventually, old oceans will disappear completely. Great mountain chains like the Alps and the Himalayas formed when oceans disappeared and the continents on each side collided.

Question

d Which plate did the Indian plate crash into to form the Himalayas?

Key points

- Scientists used to think that the Earth's surface features were the result of the crust shrinking as the Earth cooled down.
- The Earth's surface consists of tectonic plates, carrying continents, which move by convection currents.
- When the plates crash into each other, they force up new mountains.

Tsunami shock waves

On Boxing Day 2004, just off Sumatra in the Indian Ocean, the Earth moved. The crust of the Earth shattered and slipped, triggering a massive earthquake. It was the most powerful quake for 40 years. It gave out a shock wave as powerful as 1000 nuclear bombs.

The Earth's crust moved along the boundary between the Indian and Eurasian plates. One side was lifted by 10 m – and so was the ocean above it. The water slumped back creating a giant wave – a **tsunami** – which spread across the ocean as fast as a jet plane. In deep water the tsunami was not a problem. But as it approached the shore the moving water bunched up, raising the wave into a solid wall of water up to 10 m high.

From the shore the first sign of trouble was a sudden drop in sea level. It was as if the plug had been pulled from the ocean. People walked out to look at the fish stranded on the sand, the suddenly exposed rock pools and the reefs. Then there was a roar in the distance as the tsunami approached. Within hours 250 000 people had been killed.

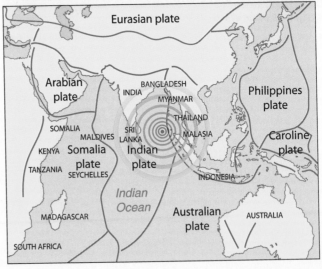

Every now and then the news brings horror stories of a devastating earthquake or an erupting volcano that engulfs land, cities and people. Many powerful earthquakes happen every year. Most of these centre on uninhabited regions and so cause few problems, but in 1999 a terrible earthquake struck Izmit in Turkey, killing 40 000 people. In 1976 an earthquake 10 times as powerful killed 600 000 people in China. Volcanoes can also be deadly. The eruption of Mount Pinatubo in 1991 in the Philippines killed 350 people and 25 000 people died after the eruption of Nevado del Ruiz in Columbia in 1985.

Where do quakes and eruptions occur?

Earth movements cause sudden and disastrous earthquakes and volcanoes. Powerful earthquakes only occur along plate boundaries. The destructive ones occur where the plates are crashing into each other like cars colliding head-on and crumpling or sliding sideways past each other.

This happens:

- along the 'ring of fire' around the Pacific Ocean where one plate is slipping down under another and being destroyed
- along the Alps and Himalayan belts where new fold mountains are created as plates collide
- in zones such as the San Andreas fault zone in California where plates slip sideways
- in the middle of the Atlantic and Pacific Oceans where new ocean crust is forming as plates move apart

Questions

a Why doesn't Britain get powerful earthquakes?
b Which region of the USA has active volcanoes, the Atlantic coast or the Pacific coast?

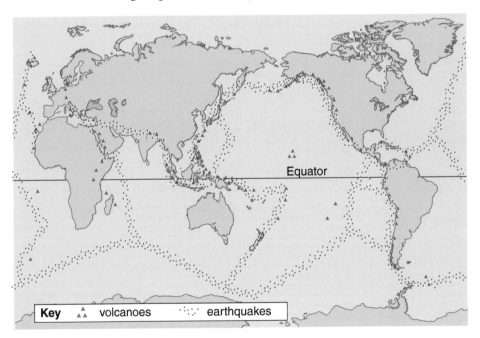

Key   ▲▲ volcanoes   ∴∴∴ earthquakes

Question

c What part of Britain is closest to the edge of the plate?

Away from the edge

We are lucky in Britain because we are a long way from the edge of the Eurasian plate. Although we don't have powerful earthquakes, there were small quakes in Wales in 1984 and 1990, and one in the Midlands in 2000. People usually report feeling the ground or buildings shake as though a large lorry had passed in the street. There are no live volcanoes in Britain now, but Arthur's seat in Edinburgh was an active volcano 400 million years ago and vast amounts of lava poured out from volcanoes along the west coast of Scotland just 50 million years ago.

Key point

- Tectonic plate movements can be sudden and disastrous, causing earthquakes and volcanoes at plate boundaries.

Vesuvius eruptions

The great volcanic cone of Vesuvius looms over the city of Naples in Italy. Nearly 2000 years ago, a huge eruption sent a cloud of red-hot gas down on the city of Pompeii. It killed 5000 people in minutes.

Vesuvius last erupted in 1944, when lava stopped just on the edge of the city. It has been sleeping since then, but it has a long history of eruptions and will erupt again. The only question for the people of Naples is 'When?'

When will it happen?

The people of Naples need to be given time to get out of the city. Scientists must try to predict when a volcano will blow or an earthquake will shake the earth. It is not easy, but there are some warning signs.

> ### Question
>
> **a** What was different about the 1944 eruption compared to the one that destroyed Pompeii?

> ### Question
>
> **b** A harbour near Vesuvius has risen 4m out of the sea in the last 30 years. Why is this cause for concern?

For volcanoes:

Increased temperature There is a great chamber full of molten magma beneath a volcano. Hot gases escape into the crater. If the gas gets hotter it could be a warning.

Earthquakes Hundreds of small earthquakes may occur beforehand. If it is the most active period since the last eruption, perhaps that is a sign.

Rising land As magma pushes in below the volcano, the ground rises.

For earthquakes:

Animal behaviour Local tales often talk of animals behaving strangely just before an earthquake: dogs start to bark or all the birds go silent. In the Sumatran earthquake, elephants stampeded inland to safety.

Pre-shocks Minor shocks often occur before a big shock. Perhaps these are what make the animals behave strangely.

Changes in the water levels in wells This often seems to drop in the period before an earthquake.

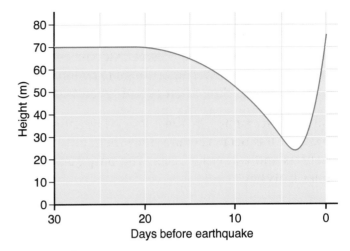

Time to evacuate?

Will the eruption or earthquake happen tomorrow, in a few months
or not for years? The best way to save lives is to evacuate people from
the danger zones. But if you do this too soon people will get bored and
drift back to their homes.

Scientists are working hard to find ways of making precise predictions
about eruptions, but it is not easy. Some volcanic craters remain open,
so it is relatively easy to monitor what is happening. They can see
whether the lava level is rising or falling, whether it is getting hotter
or cooling down and whether there is more or less gas. But with many
of the more dangerous volcanoes lava sets solid in the vent, plugging
it and stopping gas escaping. Building pressure eventually blows this
out in a massive eruption, but it is hard to tell *how much* pressure is
building up inside and *how strong* the lava 'plug' is. There are many
thousands of volcanoes in the world. Some are near large population
centres in rich countries and so are monitored carefully, but poorer
countries do not have the resources to do this.

For earthquakes, as you have seen, there are a lot of plate edges and
ocean ridges to monitor. Even if you know how they move they don't
always move steadily and the stress points may be deep underground
or at the surface. Since the Sumatran tsunami scientists are considering
installing an early warning system for tsunamis so people can at least
be evacuated from low-lying land.

There is a random unpredictability about many natural phenomena.
Some people think it will never be possible to accurately predict the
time of an eruption.

Next Vesuvius eruption?

Vesuvius is a very dangerous volcano. The old lava has turned to rock
and has plugged the crater. The pressure builds up and up inside until
the plug 'blows'. The longer it is between eruptions, the bigger the
bang when the volcano erupts. When Vesuvius last erupted, in 1944,
the pressure had had just 15 years to build up.

Year of eruption	Years since last eruption
1794	34
1858	64
1872	14
1906	34
1929	23
1944	15
20??	

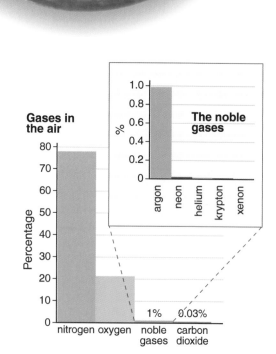

The blue and white planet

The swirls of blue and white weather patterns look fantastic. The blanket of air that covers the Earth is called the **atmosphere**. Without it, we could not exist. The atmosphere consists of many gases.

What's in the air?

The Earth's atmosphere has been more or less the same for 200 million years. 21% of dry air is oxygen, the gas you need to breathe. 78% is the unreactive gas nitrogen. There is just 1% of other gases. Normal air also contains varying amounts of water vapour. The 21% of oxygen is very important to animals on the Earth. The carbon dioxide is very important to plants.

Question

a (i) Why is oxygen essential to all animals and humans?
(ii) Carbon dioxide makes up only 0.03% of air. Why is carbon dioxide important?

What are the other gases?

A hundred years ago, nobody knew that these gases existed. Then scientists found there was just 1% of air that they could not get to react with anything. They called the gases they had discovered the **noble gases**.

Question

b Which noble gas is the most common?
c What are the names of the noble gases with the symbols He, Ne, Ar and Kr?

The noble gases form Group 0 of the periodic table. The noble gases are typical non-metals in many ways:

- they have very low melting/boiling points
- when solid they are soft and crumbly
- they do not conduct electricity
- they are very unreactive.

You may think that an unreactive gas would not be of much use. But sometimes their very inactivity is just what is needed.

Gases in the air

The noble gases

Percentage — nitrogen, oxygen, noble gases (1%), carbon dioxide (0.03%)

The noble gases (%): argon, neon, helium, krypton, xenon

Group 0

He
Ne
Ar
Kr

Using the noble gases

Helium gas is much less dense than air. It can be used in balloons – and modern airships. Hydrogen used to be used for airships before helium but it had one big drawback. It used to catch fire! Helium is used in modern airships because it is so unreactive and does not burn.

Question

d Why is helium a better gas to use in airships than hydrogen? (Hint: look at the photo!)

Neon glows if an electric current is passed through it. These discharge tubes can be coloured, making them ideal for flashy neon signs.

Argon is the cheapest of the noble gases to produce as it makes up 1% of the air. Argon is used inside ordinary light bulbs to stop the filament from burning. It is also used to give an unreactive atmosphere for welding, which could be dangerous if done in air.

Question

e What would happen to a light bulb filament if the bulb was full of air?

Krypton is used in some lasers. They are used for laser surgery and for removing birthmarks and tattoos, as well as for 'laser sights' on rifles.

Question

f Which noble gas do you use at home?

▲ The Hindenburg airship disaster.

Key points

- For 200 million years the Earth's atmosphere has consisted of 78% nitrogen and 21% oxygen. About 1% is carbon dioxide, water vapour and noble gases.
- Noble gases are in Group 0 of the periodic table and are chemically unreactive, which can be a useful property.

The first atmosphere

The Earth formed 4½ billion years ago. In the beginning things were very different from our present atmosphere of oxygen and nitrogen. The Earth was very hot and was covered with volcanoes. Gas from these volcanoes formed the first atmosphere. This first atmosphere was thought to be made mostly of carbon dioxide and water vapour, with a little methane and ammonia. Scientists have come up with this theory about the origin of the atmosphere by studying the gases from modern volcanoes.

As the Earth cooled, the water vapour condensed, forming the oceans. The atmosphere was almost 100% carbon dioxide, just like on Mars and Venus today. There was no oxygen.

Simple microbes lived in this oxygen-free environment. Then about 3 billion years ago simple plants evolved in the oceans. These plants changed the world. Fossils show they became bigger and more sophisticated. Eventually, they colonised the land as well.

The first pollutant on Earth

Plants make oxygen when they photosynthesise. To the simple microbes, oxygen was a poison so the growth and spread of plants led to the pollution of the world's oceans with oxygen because oxygen dissolves in water.

By 2 billion years ago, early life-forms were nearly wiped out. Just a few survived in oxygen-free deep ocean mud or stagnant pools.

▲ The Earth 4.5 billion years ago may have looked like this.

▲ 3 billion years ago.

▲ 2 billion years ago.

Questions

a What would happen to these primitive microbes if you stirred up the mud and let the oxygen in?

b Now and then researchers find microbes that live off methane and ammonia deep in ocean vents. How old must these kinds of organism be?

Locking up the carbon

As plants evolved and spread, oxygen started to build up in the air. Just under 1 billion years ago, the first animals appeared in the oceans. Some of these built shells from calcium carbonate. They needed carbon dioxide for this. When they died their shells formed limestone, which covered large areas of the surface of the Earth. This started to take carbon dioxide out of the air.

Meanwhile plants were also taking in carbon dioxide. More carbon dioxide was removed from the air and laid down in the rocks as fossil fuels, coal and oil. By 300 million years ago great swamp forests covered much of what is now Britain. You can find the stumps of fossil trees in many parts of the country.

Question

c *What are the two ways in which early plants changed the air to make it more like it is today?*

▲ Limestone and coal: this is where a lot of the carbon dioxide from the Earth's original atmosphere went.

Carbon dioxide or oxygen?

Look at the graph below, which shows the percentage of carbon dioxide and oxygen in the atmosphere over time.

Question

d *(i) How long did it take for the carbon dioxide level to drop to 50%?*
(ii) For approximately how many millions of years has there been more oxygen than carbon dioxide in the atmosphere?
(iii) What major gas is not shown on the graph?

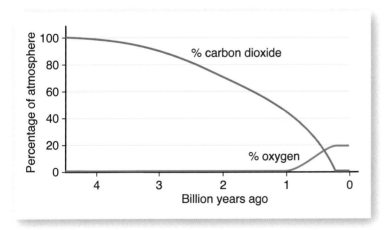

Key points

- In the first billion years of the Earth's existence the atmosphere was very different from today.
- Theories suggest that the atmosphere was mainly carbon dioxide until plants developed and produced oxygen.
- Most of the carbon dioxide from the early atmosphere is now locked up in fossil fuels and limestone rock.

Out of balance?

For 200 million years our atmosphere has been in balance. Plants take in carbon dioxide and animals breathe it out. Recycling at its best! Over the last 200 years we have upset the balance. We've been burning up the fossil fuels a million times faster than they took to form. What effect will this have on the Earth?

The debate rages

Most scientists agree that the climate is changing and that the billions of tonnes of carbon dioxide that we put into the air every year when we burn fossil fuels is contributing to this change, but it may not be the only cause. Beyond that, there is little agreement. There are several different theories about what effect human activity is having on the Earth's atmosphere now.

Carbon dioxide to blame

Environmentalists believe the amount of carbon dioxide produced is largely to blame and that we need to try to stop the climate from changing. In a meeting in Kyoto in 1997, leaders of the developed world, including Britain, agreed to set targets for reducing the amount of carbon dioxide they produce.

The biggest polluter of all, the USA, has not agreed to this. They say using less oil could harm their economy and people would suffer. They also point out that India and China, both countries with huge populations, are industrialising fast using fossil fuels and will soon produce more carbon dioxide than even the USA.

Question

a *How could India and China wreck the Kyoto agreement?*

Natural climate change

Environmentalists often give the impression that the climate would stay the same if it were not for us. Other scientists point to changes in the climate in the past. About 400 years ago, Britain was so cold in winter that the River Thames froze over in London and fairs were held on it. About 800 years ago it was much warmer than today and grapes were grown to make wine in northern Britain.

Going further back, 20 000 years ago, the world was in the grip of an Ice Age. Britain was covered by great ice sheets. A lot of water was locked up in the ice. Then 10 000 years ago the ice started to melt and sea levels rose, flooding communities that lived in low-lying areas.

Research shows that the Earth's climate has been alternating between ice age and much warmer periods every 100 000 years or so for millions of years. Our climate is currently about halfway between the two extremes. Some scientists believe this kind of climate change is natural and that perhaps human activity only plays a small part.

Plotting the changes

The graph shows carbon dioxide levels in the atmosphere in the last 2000 years.

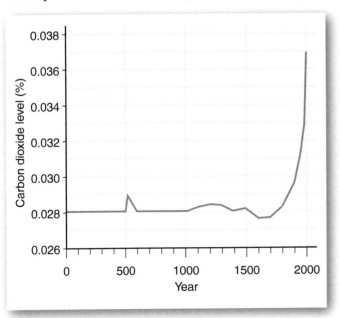

Question

b *From what you have read and heard in the news, who or what do you think is responsible for our changing climate?*

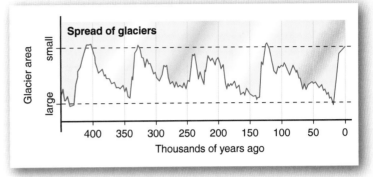

Question

c *Why is it unlikely that the Earth's climate will stay constant?*

Question

d *(i) Describe in words what the graph shows.*
(ii) For roughly how long has the recent rise in carbon dioxide levels been going on?
(iii) What human activity over the last 200 years might have caused this?
(iv) A huge volcano erupted in 500 AD. What effect did this have on the atmosphere?

Key points

● We are releasing carbon dioxide locked up in fossil fuels from the Earth's early atmosphere and this is increasing CO_2 levels in today's atmosphere.
● We can evaluate theories about the changes occurring in today's atmosphere by looking at the evidence.
● We can evaluate the effects of human activity on today's climate by analysing data.

1 Choose words from the list for each of the spaces **1–4**.

    A cracking   **B** alkene
    C temperatures   **D** hydrocarbon

The fractional distillation of crude oil produces more large ____**1**____ molecules than are needed. Fortunately these can be chopped up into smaller pieces by a process called ____**2**____. This happens when the hydrocarbons are heated at a very high ____**3**____. Cracking makes more petrol but also small ____**4**____ molecules used to make plastics.

2 Plastics (polymers) such as poly(ethene)are made from crude oil. Four stages in the process are the following:

    i   very long-chain molecules (polymers) are formed
    ii   small molecules are formed
    iii large molecules are cracked
    iv small molecules are made to 'pop' together

Which of these orders is correct?

    A iv → ii → i → iii
    B iii → ii → iv → i
    C ii → i → iv → iii
    D i → ii → iii → iv

3

	Polymer	Properties
1	poly(propene)	semi-rigid, easily coloured
2	poly(ethene)	cheap, flexible
3	poly(styrene)	rigid, easily moulded
4	PVC	tough, resists damage, shiny

Which polymer, **1–4**, would you use

    a to make a supermarket carrier bag   *(1 mark)*
    b to make the case for a CD player   *(1 mark)*
    c as 'artificial leather'   *(1 mark)*
    d to make a washing-up bowl?   *(1 mark)*

4 The graph shows how much energy you get by burning 1 kg of plastic waste compared to 1 kg of oil and coal.

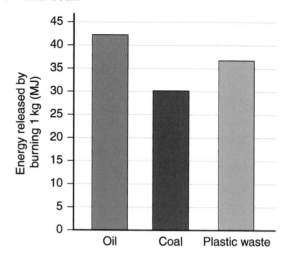

  a The graph shows that
    A It would be a great waste to burn plastic.
    B Plastics produce a similar amount of energy per kg when burnt as coal and oil.
    C Burning plastic would cause pollution.
    D 1 kg of plastic gives 30 MJ of energy when it is burnt.

  b Most waste plastic in Britain ends up in landfill sites. This is not good because
    A Plastics are biodegradeable.
    B Seagulls come down and eat the plastic.
    C It is a waste of a valuable resource.
    D Plastic bags make the landfill sites look untidy.

  c It might be a good idea to burn plastic waste in a power station because
    A We would have to use less fossil fuel.
    B It would make less carbon dioxide.
    C We would get cheap electricity.
    D It would be a cleaner fuel than oil.

  d Some plastics contain nitrogen or chlorine. These might cause problems because
    A They are explosive gases.
    B They do not burn well.
    C They could help make acid rain or smog when they burn.
    D They reduce the energy output.

5 Choose words from the list for each of the spaces **1–4**.

    A decolourise
    B unsaturated
    C energy
    D saturated

Many seeds contain oil as an ____**1**____ store for their growing seedlings. ____**2**____ oils have only single C–C bonds. ____**3**____ oils have at least one double C=C bond and are able to ____**4**____ bromine water.

6 Match words **A**, **B**, **C** and **D** to the sentences in the table.

    A whipped cream   **B** egg yolk
    C cream   **D** butter

1	This is a 'water' in oil emulsion made by churning milk.
2	This oil in water emulsion separates out from milk.
3	This oil in water emulsion has added air bubbles.
4	This is added to oil and vinegar in mayonnaise as an emulsifier.

7 a Choose the correct word or phrase from each pair to complete these two sentences.

    i Animal fats are said to be (**saturated/unsaturated**) as they have as many hydrogen atoms as possible, with just single carbon to carbon bonds. *(1 mark)*

    ii Vegetable oils have one or more C=C double bonds and are said to be (**saturated/unsaturated**). *(1 mark)*

b Which type of fat/oil is thought to be better for your health? *(1 mark)*

c Which gas is bubbled through hot vegetable oils to turn it into solid fat for margarine? *(1 mark)*

d Why is nickel used in this process? *(1 mark)*

e What is this process called? *(1 mark)*

8

Fuel	Energy content (MJ/litre)	Source	Main pollutants produced compared to petrol engine *		
			Particulates	Nitrogen oxide	Carbon monoxide
petrol	26	crude oil	*	*	*
diesel	29	crude oil	high	*	*
biodiesel	27	vegetable oil	low	high	*
ethanol	18	sweetcorn	*	low	low

a How does biodiesel compare to petrol and diesel in terms of energy content per litre? *(1 mark)*

b How does ethanol compare to petrol and diesel in terms of energy content per litre? *(1 mark)*

c Cars can be converted to run on either petrol or ethanol. Which do you think would give more miles per litre? *(1 mark)*

d City smog is caused by a mixture of nitrogen oxides, carbon monoxide and particulates. Is ethanol more or less likely to cause city smog? *(1 mark)*

e What pollution disadvantage does biodiesel have compared to ethanol? *(1 mark)*

f What big advantage do biodiesel and ethanol have over petrol and diesel in the long term? *(1 mark)*

9 Choose words from the list for each of the spaces **1–4**.

    A earthquakes   **B** wrinkles
    C currents   **D** plates

Scientists once thought that mountains were like ____**1**____ that formed as the Earth cooled and shrank. We now know that the crust of the Earth is broken up into slabs called ____**2**____. These move slowly due to convection ____**3**____ in the hot rocks beneath them. Along their edges the rocks can 'stick' and then jerk suddenly causing powerful ____**4**____.

10 Yellowstone Park is a vast area in the Rocky Mountains of Wyoming, USA. It is famous for its geysers that squirt boiling water into the air in great fountains. It also has bubbling pools of mud and sulfurous fumaroles. Recently there have been many small earthquakes, and the ground has risen by a metre or so in some areas.

a Give **two** features that suggest that Yellowstone is a volcanic region. *(2 marks)*

b Give **two** features that suggest an eruption might be on the way. *(2 marks)*

11 This question is about the atmosphere and how it has changed over the earth's history.

a The composition of the air today is approximately
   A 1/5th nitrogen and 4/5ths oxygen
   B 1/5th carbon dioxide and 4/5ths oxygen
   C 1/5th oxygen and 4/5ths carbon dioxide
   D 1/5th oxygen and 4/5ths nitrogen.

b Four billion years ago the Earth's atmosphere was mostly made from
   A nitrogen
   B oxygen
   C methane
   D carbon dioxide.

c There is oxygen in the atmosphere because
   A Animals need oxygen to breathe.
   B Animals make oxygen when they respire.
   C Plants make oxygen during photosynthesis.
   D Methane reacts with carbon dioxide.

d There is less carbon dioxide in the atmosphere on Earth than there once was because
   A We burn fossil fuels on Earth.
   B Lots of carbon is locked up in limestone and fossil fuels.
   C Animals breathe out carbon dioxide.
   D It has turned into nitrogen.

e Just under 1% of the air is a gas called argon. This gas was not discovered for a very long time because
   A It is invisible.
   B It just has single atoms.
   C It is an unreactive gas from Group 0 of the periodic table.
   D It is very similar to oxygen so it was very hard to tell the two gases apart.

12 a i Over the last 200 years we have been burning fossils fuels at a faster and faster rate. What effect has this had on the composition of the atmosphere?
   (1 mark)

   ii Why are some people worried about this change? *(1 mark)*

   iii What can we do to reduce this effect?
   (1 mark)

b China is building lots of new coal-fired power stations. Why is this a concern? *(1 mark)*

c The climate does seem to be changing. What evidence is there that suggests it might not be due to burning fossils fuels alone? *(2 marks)*

d What natural event can put vast amounts of carbon dioxide into the atmosphere? *(1 mark)*

13

Location	Type of building
England (traditional)	brick and mortar
Japan (traditional)	wood and paper
USA (California – modern)	concrete and steel

a Which of the regions shown above suffer from powerful earthquakes? *(2 marks)*

b i During an earthquake the ground shakes violently. What would happen to a traditional English house in a powerful earthquake? *(1 mark)*

   ii Why would this be very dangerous to people in the house? *(1 mark)*

c i What would happen to a traditional Japanese house in a powerful earthquake? *(1 mark)*

   ii Why would this be less dangerous to people in the house compared to an English-style house? *(1 mark)*

d i Concrete and steel-framed buildings are flexible and can sway without breaking. Are skyscrapers more or less vulnerable than 'bricks and mortar'? *(1 mark)*

   ii Which material used extensively in modern skyscrapers might be more vulnerable to earthquake damage? *(1 mark)*

14 Chloroethene is a chemical with the formula C_2H_3Cl. It used to be called vinyl chloride.

  a This compound has a C=C double bond. What part of its 'new' name tells you that? *(1 mark)*

  b Chloroethene is a monomer that can be made to polymerise. What is the correct name for this polymer? *(1 mark)*

  c What is this polymer usually called (based on the monomer's 'old' name)? *(1 mark)*

  d This polymer is often used to coat electric cables. What property makes it ideal for this use? *(1 mark)*

  e This polymer is also used to make the frames for many modern windows. What property makes it a very good material for this? *(1 mark)*

  f Most monomers are hydrocarbons. Is chloroethene a hydrocarbon? If not, why not? *(2 marks)*

15 Suggest a reason for each of the following:

  a Green dye is added to processed peas. *(1 mark)*

  b Onions are kept in jars full of acetic acid. *(1 mark)*

  c Monosodium glutamate is added to Chinese food. *(1 mark)*

  d A small amount of sugar is added to baked beans. *(1 mark)*

  e Large amounts of sugar are added to jam. *(1 mark)*

  f Red dye is added to tomato ketchup. *(1 mark)*

16 a Which of the following are examples of emulsions?

    Flora margarine, butter, lard, olive oil, milk, orange juice. *(1 mark)*

  b Mayonnaise is a thick emulsion. What advantage does this have over oil and vinegar salad dressing? *(1 mark)*

  c What advantage does emulsion paint have over other paints (especially when painting the ceiling!). *(1 mark)*

  d What happens to all emulsions if they are left for long enough? *(1 mark)*

  e Whipped cream has a more complex structure than double cream. What is the 'extra feature' in its emulsion? *(1 mark)*

17 This drawing shows some of the continents as they were 200 million years ago.

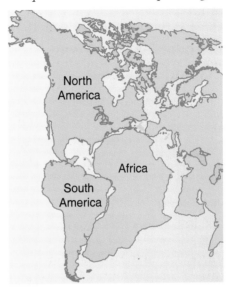

  a What is the modern name that is given to the process that has moved them apart? *(1 mark)*

  b What name did Alfred Wegener give to this process? *(1 mark)*

  c What did scientists think about this idea when Wegener first suggested it in 1912? *(1 mark)*

  d What did scientists discover running along the middle of the Atlantic Ocean that helped to make them change their minds? *(1 mark)*

  e Explain how the Atlantic Ocean is still getting wider, year by year. *(1 mark)*

  f If the Atlantic Ocean is getting wider, what must be happening to the Pacific Ocean, on the other side of the planet? *(1 mark)*

  g Explain why there is a 'Ring of Fire' around the Pacific Ocean. *(1 mark)*

Keeping warm

When we heat a house, we want it to warm up quickly and cheaply. We need to transfer heat effectively through the building. Once it is warm, we want it to stay that way, so we need to stop the heat leaking away. To use energy efficiently, we must understand how heat is produced and transferred.

▲ It's nice and cosy in here...

▲ ...but not so warm in here!

Heating our homes

Heating our homes takes a lot of energy. This is expensive, and much of the heat is wasted. About 30% of the carbon dioxide produced in the UK comes from energy used in our homes. Carbon dioxide contributes to global warming, so it is important to reduce the amount we produce. If we know how **thermal energy** (heat) is transferred from one place to another, we can try to reduce the amount that is wasted. Houses of the future will need to use less energy or use it more efficiently.

Imagining homes of the future

Some people, like Bill Gates of Microsoft, have decided to build their houses mainly underground. This keeps thermal energy in the house and stops so much of it escaping into the air.

The American TV star Julia Louis-Dreyfus and her husband Brad Hall have adapted their house in California to incorporate many energy-saving features.

Part of the roof opens in hot weather. The hot air inside can escape, drawing cool ocean breezes in through the windows. This means that the house does not need air-conditioning.

Solar heat panels on the roof produce hot water for the house. The panels contain water pipes. They are heated by the sun and the thermal energy is **transferred** (moved) through the walls of the pipes into the water.

The house is designed with many large windows so hardly any electric lighting is needed during the day. When electric lights are used, they have low-energy light bulbs. The electricity comes from solar cells on the roof that convert light into electricity.

▲ One of Bill Gates' houses – much of it is underground.

▲ The houses on this estate in Milton Keynes are installed with solar panels and other energy saving devices.

Think about what you will find out in this section

How is thermal energy transferred from one place to another?	How can we conserve energy in our homes?
What factors affect the rate at which thermal energy is transferred?	How can we become more cost-effective in our use of energy?
What is meant by the efficient use of energy?	

Cooking with conduction

If you pick up a saucepan that has a metal handle, you may burn your hand. Thermal energy has been transferred along the handle from the hot saucepan to your hand.

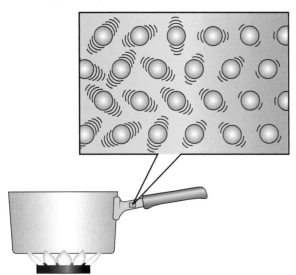

▲ The metal part of the handle of this saucepan conducts thermal energy easily.

The particles in the end of the handle joined to the hot saucepan start to vibrate vigorously as they warm up. They pass vibrations to their neighbours. This is called **conduction**. Gases and liquids are worse than solids at conducting thermal energy. Their particles move around and have no fixed neighbours so it is hard for them to transfer the vibrations to each other.

Question

a Some saucepans have holes cut in the metal handles, like the one shown here. How do the holes make the handle cooler to pick up?

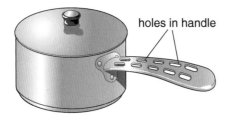

holes in handle

Some solids are much better at conducting thermal energy than others. Metals like copper and silver are excellent **conductors**. Many ceramic materials, and plastics such as polystyrene, are good **insulators**. Good insulators are very poor conductors.

Questions

b Why are high quality saucepans often made with copper bottoms?
c Why are disposable cups often made from polystyrene?

Heating our homes with convection

In central heating systems, a boiler heats water, which is then pumped around to radiators. Air near the radiator becomes warm. Air is a mixture of gases. When heated, the particles of gas in the air gain energy and vibrate faster. This makes them spread out more than before.

Because the particles are further apart in the warm air, it is less dense than the cool air in the rest of the room. Less dense substances float on denser substances so the warm air rises and the cooler air sinks to take its place. This flow of air is called **convection**. It creates a circulating **convection current**.

Hot showers from convection

Convection is used in this hot water system. The boiler downstairs heats the water. The warm water is less dense so it rises up into the hot water tank. Cooler, denser water falls to take its place. Because of convection, there is no need to pump the water around.

Convection is efficient when particles can flow freely, as they can in air or water. Thick, syrupy liquids do not transfer thermal energy so well by convection. Convection does not happen at all in solids.

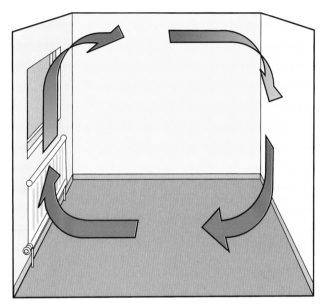

▲ A room is warmed by convection.

▼ A hot water system.

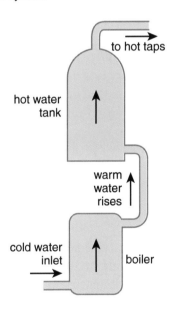

Questions

d Why is the outlet for the hot taps put at the top of the hot water tank rather than at the bottom?

e Why can convection take place in gases and liquids but not solids?

f Look back at the description of Julia Louis-Dreyfus' house. List one energy-saving feature that depends on conduction and one that depends on convection. Explain your answers.

Key points

- In conduction and convection, thermal energy is transferred by particles.
- In conduction, the thermal energy moves from place to place but the particles do not. In convection, the particles move.
- Some materials transfer thermal energy faster than others.

Heat without particles

Thermal energy from the Sun reaches us across 150 million kilometres of empty space.

There are no particles in space to transfer thermal energy, so the Sun's energy cannot be arriving by conduction or convection. Conduction and convection cannot happen in a vacuum.

Energy from the Sun is transferred to Earth by **radiation**. Radiation transfers energy without particles. Thermal or **infra red** radiation is an **electromagnetic wave**, rather like light energy. It is a form of energy and can travel through a vacuum. Unlike light energy, it is invisible to the human eye.

When you sit in front of a fire, you can feel the heat even from far away. Thermal radiation travels to you through the air. Radiation from an electric grill also travels through the air – there is no contact between the grill and the food being cooked.

In these cases, there are air particles in the way of the radiation but the particles do not transfer the thermal energy. The radiation goes through them, rather like light through a window.

▲ Thermal energy reaches us from the Sun.

▲ The electric grill transfers energy by thermal radiation.

> **Question**
>
> a How is the thermal energy from a toaster transferred to the bread?

Absorbing and emitting radiation

When you sit in the sun, you take in or **absorb** thermal radiation but your body is itself warm, and you also give out or **emit** thermal radiation to the surroundings. In fact, every object absorbs and emits radiation all the time. The hotter an object is, the more radiation it emits.

> **Question**
>
> b Which emits more thermal radiation, a glass of cold water or a cup of hot tea?

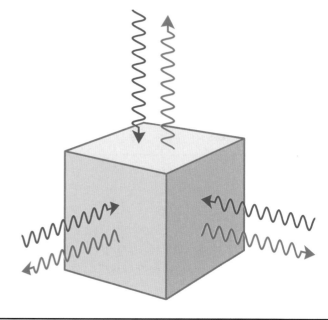
▶ Every object is absorbing and emitting thermal energy all the time.

Surfaces, radiation and reflection

Some surfaces are good absorbers of radiation. When radiation hits them, most of it is absorbed. The rest simply bounces off – it is reflected. Other surfaces are bad absorbers. When radiation hits them, only a little is absorbed and most of it is reflected.

Two jars of cold water were left in the sun. One was painted black and the other white. The water temperature was measured every minute.

The graph of the results (see left) shows that the water in the black jar heated up more quickly. The black jar must be better at absorbing thermal radiation. The white jar absorbed less radiation; instead, it reflected radiation.

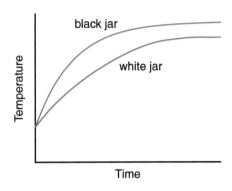

Questions

c Which are better absorbers of thermal radiation, black or white surfaces?
d On a sunny day, do the black stripes or the white stripes of a zebra crossing get hotter? Why?

Surfaces that are good at absorbing radiation are also good at emitting it. The black and white jars were filled with hot water and left to cool down. The water in the black jar cooled more quickly. The black jar is better at emitting thermal radiation, just as it is better at absorbing it.

Question

e Which are better emitters of thermal radiation, black or white surfaces?

The texture of a surface also affects how good it is at absorbing and emitting radiation. Matt surfaces, such as cork or wood, are good absorbers and emitters. Shiny surfaces, like mirrors, are poor absorbers and emitters. They reflect radiation instead of absorbing it.

Question

f Which is a better absorber of radiation, the mirror or the wall? Which is a better emitter?

Key points

- Thermal radiation transfers energy by electromagnetic waves without the need for particles.
- Dark surfaces are good absorbers and emitters of thermal radiation.
- Light or shiny surfaces are poor absorbers and emitters.
- The radiation that is not absorbed by a surface is reflected.

Size, shape and energy transfer

The size of an object and the size of its surface are important for thermal energy transfer.

Conduction depends on particles passing vibration to their neighbours. A big piece of copper has more particles than a small piece, so it conducts thermal energy faster. Large objects conduct at a greater rate than small objects made from the same material.

> **Question**
>
> **a** Two cups of tea are on a table. Someone puts a small teaspoon in one cup and a much larger tablespoon in the other. Which spoon conducts thermal energy out of the tea faster? Why?

For radiation, what matters is the object's surface because objects absorb and emit radiation only at their surfaces. The bigger the surface, the faster it absorbs and emits radiation. An object with a large surface area absorbs and emits radiation at a greater rate than an object with a small surface area.

Solar heating panels absorb energy from the Sun, and transfer this energy to water circulating in pipes below the panels.

> **Questions**
>
> **b** How does the surface area of the solar panels affect the rate of absorption of energy from the Sun?
> **c** Which panel will emit thermal radiation to the air at a faster rate – the large or small surface area?

Shape is also important for thermal energy transfer. Convection needs a free flow of gas or liquid. Gaps in or around an object let the gas or liquid flow easily. An enclosed shape makes this harder.

> **Question**
>
> **d** A homeowner builds a wooden cover around a radiator. Explain why this makes the radiator less good at warming the room.

Cooling your food, heating the room

A refrigerator moves thermal energy out of your food to keep it cool. The energy is transferred to the metal grille at the back of the refrigerator – that's why a fridge or freezer can warm up the room as well as keeping the food inside cold.

The grille loses thermal energy to the air by radiation and conduction. This makes the surrounding air warm up. The warm air rises by convection. This carries thermal energy into the rest of the room.

The grille is painted black to help it emit better. The many slats have a large surface area. The gaps between the slats let the air circulate.

Question

e *How is the grille designed to help it transfer thermal energy to the room by radiation and convection?*

Cool and colder, warm and hotter

A glass of lemonade may be cool, but it is warmer than an ice cube. When you put ice into the lemonade, thermal energy is conducted from the drink into the ice. The ice warms up – you can see this happening because as it warms up, it melts.

If you put an ice cube into a cup of tea, it warms up and melts much more quickly. There is a much bigger **temperature difference** between hot tea and ice than between cold lemonade and ice. The bigger the difference in temperature, the faster thermal energy is transferred.

Question

f *Use ideas about thermal energy transfer to explain why you get cold more quickly when swimming in the sea than when swimming in a heated pool.*

Key points

- The size and shape of an object affect the rate at which it transfers energy.
- The bigger the temperature difference between an object and its surroundings, the faster the rate of thermal energy transfer.

▲ A refrigerator grille is designed to lose thermal energy easily.

Cooking and heating

If you are cooking food or heating a room, you want to transfer thermal energy effectively from one place to another. But if you are trying to stay warm, paying the heating bill or just keeping your coffee hot, you want to reduce heat loss as much as possible. This means you need to reduce thermal energy transfer by conduction, convection and radiation.

We cannot see thermal energy being transferred but we need to ensure that thermal energy goes where we want it, and stays there.

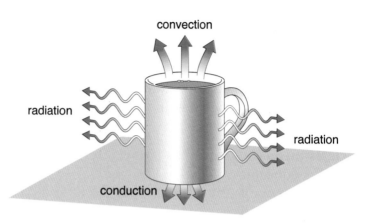

▲ Thermal energy transfer is invisible.

Keep it warm, keep it cool

An everyday example of a device designed to stop thermal energy transfer is the vacuum flask.

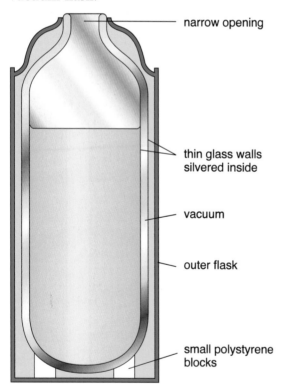

- narrow opening
- thin glass walls silvered inside
- vacuum
- outer flask
- small polystyrene blocks

The flask has a pair of thin glass walls with a vacuum between them. The glass is coated with a layer of shiny metal. The glass flask rests on small polystyrene blocks inside a protective outer flask.

▲ This person's hair, clothes and glasses reduce thermal radiation.

Questions

a How does the vacuum reduce thermal energy transfer?
b How does the shiny surface slow down thermal energy transfer?
c Why is polystyrene chosen to support the inner flask and why are small blocks used?
d Explain how the same flask can be used to keep hot drinks hot or cold drinks cold.

Choosing the right materials

We can measure how fast thermal energy is transferred through a window, wall or other building component. The rate is expressed as a U value. The lower the U value, the better the window is at stopping thermal energy transfer.

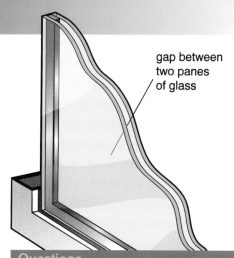

gap between
two panes
of glass

Type of window	Typical U value (W/m²/K)
A double glazing using glass coated with a poor emitter of thermal radiation; air between the layers of glass	2.1
B single glazing	5.2
C double glazing with air between the layers of glass	3.0
D double glazing using coated glass and vacuum between the layers of glass	0.8

Questions

e List the windows in order, starting with the best at reducing thermal energy transfer.
f How does each feature improve the window's thermal insulating properties?

Braving the Arctic chill

The British explorer Ann Daniels has experienced temperatures below −50 °C in the Arctic. At temperatures like that, excessive heat loss means death. The picture shows how she reduced the amount of thermal energy escaping from her body.

Question

g How does each of the features of the explorer's equipment reduce thermal energy loss? Use the three ways of transferring thermal energy in your answer.

Balaclava, hat and hood to reduce heat loss from the head

Fur trim to cover exposed skin on face and wrists

Special fabrics woven from insulating fibres

Three layers of clothing. Between them are trapped layers of still air, which cannot circulate

▲ The explorer Ann Daniels has visited both poles.

Key points

● It is possible to predict how thermal energy will be transferred into and out of objects.
● Equipment can be designed to reduce thermal energy transfer.

Transforming energy

Energy comes in many forms: thermal, chemical, kinetic, potential and so on. In order to use it, we must **transform** the form we have into the form we want. For example, coal contains chemical energy. When we use it to heat a house, we must transform that energy into thermal energy.

Transforming energy is the only way to get the energy we want because it is impossible to create energy from nothing. Similarly, it is impossible to destroy energy.

Whenever we transform energy or transfer it from place to place, some of it is wasted. We can never transform all the energy we have into the form we want.

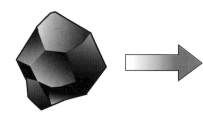

chemical energy

thermal energy

Question

a Explain why the total amount of energy in the Universe always stays the same.

Useful energy and wasted energy

A light bulb is intended to transform electrical energy into light energy, but it also produces a lot of unwanted thermal energy. The form we want – light – is **useful energy**. The form we do not want – heat – is **wasted energy**.

Which energy is 'useful' and which is 'wasted' depends on what we want. The thermal energy produced by an electric heater is useful, but the thermal energy produced by a light bulb is wasted. The sound from a loudspeaker is useful, but the sound from a hairdryer is wasted.

◀ Don't touch – it's hot!

Device	Form of input energy	Useful form of output energy	Main forms of wasted energy
light bulb	electrical	light	thermal
dynamo		electrical	thermal, sound
solar electric cell		electrical	thermal
hairdryer	electrical		
gas cooker			light

Question

b Copy and complete the table above.

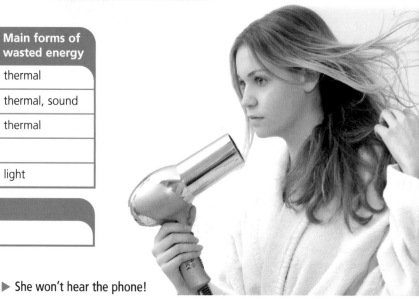

▶ She won't hear the phone!

Wastage at every step

Many devices transform energy in several steps. Each step usefully transforms only part of the energy it starts with. By the end of the whole process, we have often wasted a large proportion of the energy.

In a car, fuel is burned to produce exploding gas, which moves a piston. Only about a quarter of the chemical energy in the fuel is transformed into the kinetic energy of the piston. The rest is wasted, for example as unwanted heat and noise.

When the piston's energy is transferred to the wheels, more is wasted. Some is transformed into more noise and thermal energy generated by friction. Only about half the energy from the engine – an eighth of the chemical energy we started with – is used to move the car.

▲ Some of the energy wastage from a car.

Question

c *When fuel is used to move a car, what is the useful form of energy produced?*

At each step, the energy becomes more spread out and harder to use. The chemical energy in the petrol is concentrated in a small space. Thermal energy is spread over much of the car, and then spreads even further into the surroundings.

▲ Energy spreads out each time it is transformed.

The useful energy spreads out too. As the car moves, it generates thermal energy from friction with the air and the road. Eventually, all the energy is transferred to the surroundings as thermal energy, and the surroundings get warmer.

Question

d *A battery is used to run a toy train. When the battery is flat and the train has stopped moving, where is the energy and why isn't it useful now?*

Key points

- Energy cannot be created or destroyed.
- Whenever energy is transformed or transferred so we can use it, some energy is wasted.

Making the best of our energy

Energy does not come free. Most forms of energy cost money. There are other costs too. A lot of our energy comes from fossil fuels. Burning fossil fuels to produce heat releases carbon dioxide into the atmosphere. Many people think that this contributes to global warming. So we want to make the best of the energy we use.

What is efficiency?

The more **efficient** a device is, the more of the input energy it transforms into the form that we want and the less it wastes.

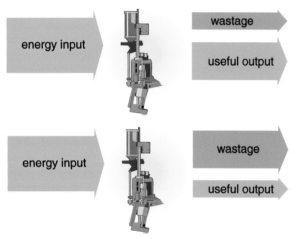

▲ The upper device is more efficient than the lower one. It transforms more energy usefully and wastes less.

For example, energy-saving light bulbs are more efficient than ordinary tungsten light bulbs. More of the electrical energy we supply is transformed into light and less is wasted as heat.

> ### Questions
>
> **a** Why is it important to know about energy efficiency?
> **b** Does an energy-saving light bulb get hotter or less hot than an ordinary tungsten light bulb when both are switched on? Explain your answer in terms of efficiency.

▲ Energy comes at a cost.

Calculating efficiency

Efficiency is calculated by dividing the useful output energy by the total input energy.

$$\text{efficiency} = \frac{\text{useful output energy}}{\text{total input energy}}$$

As an example, let's look at how efficient an ordinary light bulb is. The energy is measured in joules (J).

Total electrical energy supplied (J)	100
Amount of light energy given out (J)	5
Other forms of energy given out (J)	95

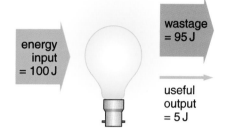

The useful output in this case is the light. Everything else is wasted energy, so we ignore it.

$$\text{efficiency} = \frac{\text{useful output energy}}{\text{total input energy}} = \frac{5\,\text{J}}{100\,\text{J}} = 0.05$$

Question

c 100 J of electrical energy are supplied to an energy-saving light bulb. It produces 25 J of light energy. What is its efficiency?

Which devices are the most efficient?

The greater the efficiency, the more efficient the device is. A perfectly efficient device would usefully transform all the energy we supplied and would not waste any, so the efficiency would be 1.

In fact, no real device is perfect – all waste some of the energy supplied, so efficiency is always less than 1.

Device	Total energy supplied	transformed
tungsten light bulb	500 J electrical energy	25 J light energy
electric fan motor	500 J electrical energy	400 J kinetic energy
solar electric cell	500 J light energy	75 J electrical energy
energy-efficient light bulb	500 J electrical energy	125 J light energy
domestic gas boiler	500 J chemical energy	425 J thermal energy

Questions

d Imagine that 250 J of electrical energy were supplied to a perfect light bulb.
 (i) How much energy would be transformed into light?
 (ii) What would be the bulb's efficiency?
e List the devices shown in the table from most to least efficient.

Key point

● Efficiency measures the proportion of input energy that is usefully transformed.

It pays to save energy

Wasted energy costs money. The less energy you waste, the more money you save. For example, the electricity required to run an ordinary light bulb for 4 hours per day for a year costs £10. The equivalent cost for an energy-saving light bulb is only £2.50.

◀ An ordinary light bulb.

▼ An energy-saving light bulb.

Questions

a Assuming a light is used for 4 hours per day, how much would you save in annual running costs by replacing an ordinary bulb with an energy-saving bulb?

b How much would a school save in running costs each year if it uses 200 bulbs for 4 hours per day?

Effectiveness and cost-effectiveness

Energy-saving measures save money, but they can also cost money. For example, if you switch to energy-efficient bulbs, you must buy new bulbs. So you need to consider what it costs to use energy-saving devices, as well as what you save.

BUY DOUBLE GLAZING!

Did you know that **20%** of the energy lost from your house goes through the windows?

That's about **£15** wasted per year.

Double glazing can slash the waste by half, to just **£7.50**.

AQA WINDOWS

Question

c How much would you save per year in wasted energy costs by switching to double glazing?

Installing double glazing saves about the same amount per year as switching an ordinary bulb for an energy-saving bulb. Both methods are equally good or **effective**. If the savings make up for the cost, we say the measure is **cost-effective**. Double glazing is more expensive, so it takes longer for the savings to make up for the cost. Switching bulbs is more quickly cost-effective because the savings make up for the cost sooner.

There is more to reducing energy consumption than cost-effectiveness. For example, double glazing decreases noise as well as reducing thermal energy loss. Some people feel strongly about making the best of energy and do not mind paying extra in order to do so.

Energy consumption – weighing the evidence

Here are some ways of reducing energy consumption in the home.

	Method	Cost	Annual saving (£)
A	run washing machine at 40°C instead of 60°C	0	10
B	lag hot water cylinder	10	15
C	install draughtproofing round doors and windows	40	10
D	insulate the loft	150	50
E	replace old fridge/freezer with energy-efficient model	250	40

> ### Questions
>
> **f** Which method gives the greatest annual saving?
> **g** Which method is the most cost-effective in the first year?
> **h** Give three more methods for reducing energy consumption that cost no money.
> **i** How long would it take for the savings from loft insulation to make up for the cost?

Some ways of reducing energy consumption are beyond the reach of individuals. Governments and businesses can help too, for example by providing energy-efficient public transport or by funding research into energy-efficient machines.

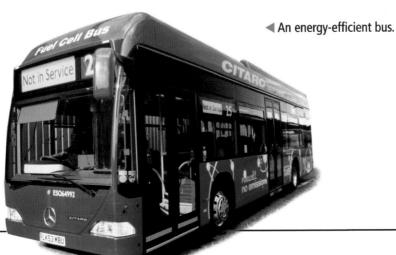

◀ An energy-efficient bus.

> ### Questions
>
> **d** Energy-saving bulbs cost more than ordinary bulbs. How does this affect the amount you can save by switching bulbs?
> **e** Energy-saving bulbs last longer than ordinary bulbs. How does this help to make up for the extra cost of buying them?

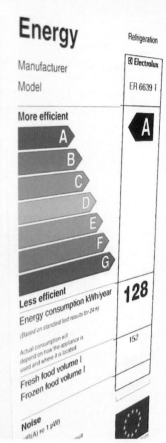

▲ By law, many domestic appliances must display labels rating their efficiency.

> ### Key point
>
> ● We can analyse the costs and savings of the devices we use and then evaluate their effectiveness and cost-effectiveness.

Every time you switch on a light, use a computer or make a phone call you are using electricity. We rely on electricity because it is so convenient and can be used for so many purposes. The challenge is to make electricity cheaply and cleanly from other sources of energy.

THE DAY THE LIGHTS WENT OUT

Where would we be without electricity?
On 28 September 2003, 56 million people across Italy found out for themselves when a huge power cut left them without power. It was the night of a carnival. Public transport was still running and cafes were open into the early hours of the morning. When the power went off, thousands of people were stranded on underground trains. Many thousands more had to sleep in the street or in train stations.

Relying on electricity

Imagine a world without electricity. No electric lights or heaters. No TV, radio or hi-fi. No computers or telephones.

We depend on electricity for light, communications and manufacturing. This means that we also depend on the fuels and other energy sources used to make electricity.

◀ Could you cope without electricity?

Fuels to generate electricity

Burning coal, gas and oil releases large amounts of thermal energy. This can be used in power stations to boil water, making steam for the turbines to generate electricity. But burning these fuels causes pollution and releases greenhouse gases, which many scientists think contribute to global warming. What's more, such fuels will not last for ever.

These are serious problems. We cannot simply stop using electricity – so it is important to look for other energy sources that are more environmentally friendly and will not run out.

Other sources of electricity

Wind, waves and sun – all provide energy from which electricity can be made.

These natural resources will not run out any time soon. Sources of energy that will last for a very long time, or that can be replaced, are called **renewable** energy sources.

At present, only about 3% of the electricity in the UK comes from renewables but the Government wants to increase that figure to 10% by 2010.

▲ Burning waste gases at an oil rig.

Think about what you will find out in this section

What is meant by power?	Why is electrical energy so useful?
How should we generate the electricity we need?	How do we calculate the cost of paying for electricity?
How can we make more use of renewable resources?	

Why are electrical devices so useful?

If you imagine trying to live without electricity, you will see how convenient electrical devices are. They can easily transform electrical energy into whatever form we want. Do you need light, sound or thermal energy? No problem, just flick the switch!

Question

a Copy and complete the table below.

Device	Input form of energy	Output form of energy
electric cooker		
computer monitor	electrical	
drill		
		light
		thermal
		sound

Electricity is also a useful form of energy because it can easily be transferred across large distances. If you want to transfer chemical energy across the country, you have to move large amounts of material such as coal or petrol. But you can transfer electrical energy through wires without moving a large amount of material.

Question

b Give two reasons why electrical devices are so convenient.

Electricity for the nation

Electricity is generated in **power stations**. The network of wires linking power stations with houses and other electricity users all over the country is called the **National Grid**. The electricity it supplies is called mains electricity.

Whenever electricity flows through a wire, the wire warms up. Some energy is wasted as thermal energy. We can reduce the wastage by reducing the current flowing in the wire.

▲ Chemical energy doesn't flow through wires!

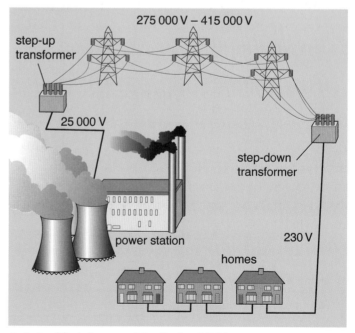

▲ A simplified diagram of the National Grid.

The higher the voltage at which the electricity is transferred, the lower the current. In the National Grid, the electricity is carried at very high voltage – up to 415 000 volts. This means that the current is low, so energy wastage is reduced. In the home electricity is used at a much lower voltage – 230 volts – because a high voltage is too dangerous.

The voltage is changed using devices called **transformers**. **Step-up transformers** increase the voltage and **step-down transformers** decrease it.

Questions

c Why is electricity transferred through the National Grid at a high voltage?
d Why is it used in houses at a much lower voltage?

▲ Guess which transforms energy faster!

Energy and power

Different devices transform energy at different rates. The greater the rate, the higher the **power** of the device. A device with higher power transforms more joules (J) of energy per second than a device with lower power.

Many electrical devices carry labels showing their power rating. The power is measured in watts (W). For example, the power of a light bulb might be 60 W. The power of an iron might be 1000 W. 1000 W is also called a kilowatt (kW). The higher the wattage, the greater the power.

Questions

e Which has a higher power: a light bulb or an iron?
f Which transforms more energy per second: a 1200 W toaster or a 2 kW heater?

The amount of energy transformed does not depend only on the power of the device. It also depends on how long the device is switched on for. A light bulb switched on all day transforms more energy than an iron switched on for 10 minutes.

▲ This device has a power of 2000 W.

Key points

- Electricity is a very convenient form of energy.
- Electricity is transferred all over the country using the National Grid.
- The power rating of a device is the rate at which it transforms energy.
- The amount of electrical energy transformed depends on how long the device is turned on for and the rate at which it transforms energy.

The right tools for the job

Both these devices transform electrical energy into light but they have quite different uses.

The desk lamp is rated at 100 W. It would not be much use in a football stadium at night. The stadium lights are rated at 2 kW each. They would be far too bright to use in a room. They would use up expensive electricity, and would produce a lot of heat as wasted energy.

◄ A desk lamp rated at 100W.

> **Question**
>
> *a Explain how each type of lamp is suited to its purpose.*

How much energy?

A stadium light transfers energy from the electricity supply at a faster rate than a desk lamp. If both are switched on for the same length of time, the stadium light transfers more energy. The longer each device is switched on, the more energy it transfers.

We can calculate how much energy a device transfers if we know its power and how long it is switched on for.

energy transferred = power × time

If we measure power in kilowatts and time in hours, then the energy is measured in kilowatt-hours (kWh). For example, if a 2 kW stadium lamp is switched on for 3 hours, then:

energy transferred (kWh) = power (kW) × time (h)
= 2 kW × 3 h
= 6 kWh

You may wonder why we do not measure the energy in joules (J). The reason is that a joule is a very small unit of energy – 6 kWh is over 20 million joules! It is inconvenient to use such large numbers, so we use kWh instead.

Don't forget that the power must be measured in kW and the time in hours. If you are given a power in watts, you must convert it to kW by dividing by 1000.

▼ Stadium lights rated at 2 kW each.

▲ It is inconvenient to use units that are too small.

> **Question**
>
> *b How much energy do the following devices transfer from the mains:*
> *(i) a 1.5 kW tumble drier switched on for 2 hours*
> *(ii) a 100 W light bulb switched on for 5 hours?*

Paying for electricity

Electricity companies charge us for each kWh of energy we use. They measure how many kWh we have used by reading the electricity meter.

total cost = number of kWh used × cost per kWh

For example, if we use 300 kWh and each kWh costs 6p, then:

total cost = 300 kWh × 6p/kWh
= 1800p or £18.00

▶ An electricity meter showing a reading of 11 483.09 kWh.

On electricity bills, a kWh of energy is called a 'unit' of electricity.

AQALec ELECTRICITY BILL

Previous reading	Current reading	Units used	Cost per unit	Total cost
45100	45500	400	7 pence	£28.00

Questions

c How did AQALec calculate the cost of £28.00?

d A 0.8 kW fan is used for a total of 5 hours each week.
  (i) How many units of electricity does it use each week?
  (ii) If electricity costs 7p per unit, what is the weekly cost?

1 September 2005

1 December 2005

▲ Meter readings taken 3 months apart.

Questions

e If electricity costs 7p per unit, what is the electricity bill for these 3 months?

f What was the average number of units used each month?

Key points

- Different devices are suitable for different applications.
- The amount of energy transferred by a device, and its cost, can be calculated.

Turning the turbine

A **turbine** is like a child's windmill. When the child blows air onto the windmill, the blades spin round.

In power stations, a giant turbine is often turned by a blast of steam at high pressure. The turbine is linked to a **generator**. When the turbine spins, so does the generator. This generates electricity.

Fossil fuels

To make steam you have to boil water. Many power stations do this by burning coal, gas or oil. These fuels were formed tens of millions of years ago from the remains of living things, so they are called **fossil fuels**.

Fossil fuels take millions of years to form. We are using them up much faster than this.

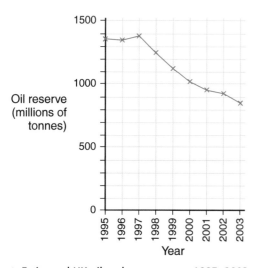
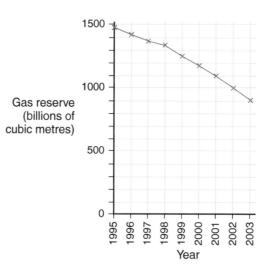

▲ Estimated UK oil and gas reserves, 1995–2003.

Questions

a What was the approximate UK oil reserve in 2000?
b Use the graph to estimate when UK gas will run out if we continue using it up at the current rate.

Choking the planet

Fossil fuels contain a lot of energy, but they have disadvantages. When they are burned, carbon dioxide is formed. Many scientists think this contributes to global warming.

Fossil fuels also cause pollution. Burning them releases acids into the air, which dissolve in rainwater to form acid rain. This can damage plants and buildings.

Gas is the cleanest fossil fuel. It causes least pollution when it is burned but it is also difficult to store as it takes up a lot of space. Coal is easier to store, but unlike gas and oil it cannot flow through pipes, so it must be transported by rail or road. Oil can be transported through pipelines, but it is very messy.

Question

c Give an advantage and a disadvantage for each of:
(i) coal (ii) oil (iii) gas.

▲ The effect of acid rain on trees.

▼ A worker in a protective suit washes radioactive waste off a building in Chernobyl.

Energy without burning

Radioactive elements like uranium and plutonium can be used in power stations in a process called **nuclear fission**, which generates thermal energy. Nuclear power stations use this energy to heat water directly.

A big advantage of nuclear power generation is that it does not release polluting gases into the air. Another advantage is that you can get a lot of energy from a small amount of material. But there is a big disadvantage. The waste from nuclear power stations is radioactive and stays that way for thousands of years. Radioactive materials damage living things and can kill people, animals and plants.

It is expensive to maintain good safety procedures while nuclear power stations are being built and when they are used. In April 1986, there was an explosion at a nuclear reactor in Chernobyl, Ukraine. A huge cloud of radioactivity was released into the air. Hundreds of thousands of people had to be evacuated. Many people have died as a result of the accident, and scientists estimate that thousands more will die from the long-term effects.

Safety procedures at nuclear power stations are meant to prevent accidents, but they were not followed at Chernobyl.

Questions

d Give an advantage and a disadvantage of nuclear power.
e Do you think nuclear power is a good alternative to fossil fuels?

Key points

- Electricity can be produced from fossil fuels or nuclear fission.
- Each method has advantages and disadvantages. Fossil fuels cause pollution and will run out. Nuclear power can be dangerous.

Spinning in the wind

We can use the wind to turn a turbine. When the wind blows, the blades spin. They turn a generator, producing electricity.

 A small wind turbine on a yacht.

Small wind turbines are used for campsites and canal boats, where there is no mains electricity. A larger turbine can generate enough electricity for 1000 homes. To make even more electricity, turbines are arranged in groups called wind farms.

The UK is the windiest country in Europe. In 2003, the Government announced the development of a number of wind farms, mainly off the coast, that would provide enough electricity for 4 million homes. The wind's energy costs nothing, and the turbines do not release pollution.

But there are disadvantages to wind power. Some people think the turbines are ugly and spoil the landscape, and the turbines only work when there is the right amount of wind. Offshore wind farms can affect shipping and may harm sea birds. Spinning turbines are noisy.

Question

a Why is the UK a suitable country for wind power?

Energy from moving water

The sea never rests. Waves constantly flow onto the shore. The tide rises and falls.

The moving water can be used to generate electricity. The world's first commercial wave power generator opened in Scotland in 2000. In this generator, the bobbing of the waves pushes air in and out of a chamber. The rushing air drives a turbine.

Twice a day, the tide carries huge amounts of water inland and out to sea again. The water can be stopped by building a dam or 'barrage'. At high tide, water builds up outside the barrage trying to get in. At low tide, water is trapped behind the barrage trying to get out. If a gate is opened in the barrage, the water flows through, turning a turbine. There is only a large build-up of water for a few hours around high tide and low tide. The rest of the time, little electricity can be made.

Wave and tide power are renewable and free but to use them we must build power stations on the seashore or on river estuaries. Barrages prevent ships from passing, and it is hard to build structures that will resist the sea's harsh waves and salty water.

▲ The tidal barrage at La Rance estuary, France.

Harnessing a river

Hydroelectric power plants also use the energy of flowing water. In most hydroelectric schemes, a dam is built across a river. This traps a huge lake of water uphill. The water has gravitational potential energy. When the dam is opened, the water rushes downhill and turns a turbine.

To create a lake behind a hydroelectric dam a huge area has to be flooded. This can devastate the environment. Forests and other habitats can be lost. Sometimes hundreds of thousands of people have to be moved away from the area before it is flooded.

Question

b Why does the trapped water have gravitational potential energy?

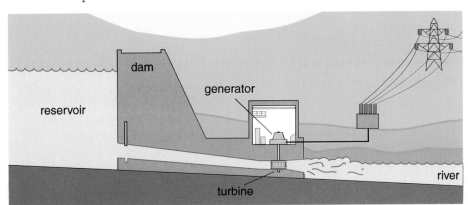

▲ A simplified diagram of a hydroelectric plant.

Question

c Give some advantages and disadvantages of wind, wave, tidal and hydroelectric schemes.

Key points

- Wind and moving water are renewable sources of energy.
- They are less polluting than fossil fuels, but have other disadvantages. Winds and tides cannot provide a constant electricity supply. Hydroelectric schemes flood large areas.

Going solar

If you have a watch or a calculator that doesn't need a battery, it probably uses a **solar electric cell**. Solar cells are made from materials that make electricity when light falls on them. They often have a battery to store the electricity for use when it is dark.

▲ A solar-powered calculator.

Question

a *What energy transformation does a solar cell carry out?*

Solar cells are silent and last a long time without maintenance. They are portable, so you can generate electricity where you need it. And sunlight is free.

▲ This medical centre in Haiti generates electricity from the Sun.

Energy from the Sun

Solar cells need light to work. They cannot generate electricity at night. And they work best in sunny countries. The UK is not very sunny, especially in winter, but even in the UK, solar cells can make some electricity.

The CIS tower in Manchester is the tallest office block in the UK outside London. It is fitted with over 7000 solar panels. Each one has an area of 0.6 m² and generates 25 kWh of electrical energy per year. The first one was fitted in July 2005.

Questions

b *How much energy will the whole tower generate per year?*

c *What area of solar panels is needed to generate 1000 kWh of energy per year?*

d *A solar panel in Germany generates 850 kWh of electrical energy per year. An identical panel in India generates 1800 kWh per year. Suggest a reason for the difference.*

A disadvantage of solar electric cells is their cost. The CIS tower project cost £5.5 million and will not even make enough electricity for the whole tower. A wind farm costing that much could generate electricity for 5000 homes.

Question

e Why might the cost of solar cells decrease if they become more popular?

Energy from the Earth

The centre of the Earth is very hot. Scientists estimate that the temperature is around 6000 °C – similar to the surface of the Sun.

As you get further from the centre, the temperature gets lower, but in some volcanic areas it is 300–400 °C just a few kilometres below the surface. Sometimes water is trapped in the hot rocks. By drilling a deep hole, we can release steam under high pressure. If no water is trapped, we can pump water in. It turns into steam and comes to the surface again. The steam can be used to drive a turbine.

Energy from the Earth's heat is called **geothermal** energy. It will not run out. Geothermal power plants do not take up much land compared with other kinds of power plant.

However, there are not very many places in the world with suitable rocks close to the surface. It is expensive drilling several kilometres into the ground, and the steam sometimes carries polluting gases up with it.

Question

f Why can't geothermal energy be used everywhere in the world?

▲ The solar panels on the CIS tower.

▲ A geothermal power plant in Iceland.

Key points

- Solar electric cells generate electricity directly from light but they are expensive and do not work in the dark.
- The Earth's heat can be harnessed to make geothermal electricity, but only a few places in the world are suitable and drilling is expensive.

Making choices

It is very difficult to decide which energy source is best for generating electricity. They all have advantages and disadvantages.

To make decisions, we must work out how much each method costs. We must consider the effect on the environment and we must think about where the electricity is needed: the best method for one place may not be the best for another place.

Building power stations

Building power stations is a major cost of producing electricity. The table shows some examples of building costs and power outputs. The power is given in megawatts (MW). A megawatt is a million watts.

It is not fair to compare the costs of power stations without thinking about how much power each one produces. The solar park in the table is cheapest, but it also makes least electricity. We need to work out how much each power station costs per MW of power that it makes.

	Type of power station					
	Gas	Coal	Nuclear	Hydro	Wind (onshore)	Solar
building cost (£ millions)	300	560	1500	140	66	35
amount of power generated (MW)	750	700	1000	100	60	10
building cost (£ millions) per MW	0.4					

Questions

a   *Copy and complete the table.*
b   *Onshore wind farms and solar parks are much faster to build than gas and coal power stations. Is it worth paying extra to get electricity quickly?*

Running power stations

Building a power station is not the only cost. We need to buy fuel, pay the workers and keep the building in good condition. To maintain the machinery it is sometimes necessary to shut it down. The start-up time afterwards varies for different types of power stations and this contributes to running costs. Nuclear power has the longest start-up time and gas-fired power the shortest.

Questions

c The costs do not take account of the pollution released by each energy source. Is this fair? Explain your answer.

d How do you think the cost of making electricity from oil and gas will change as fossil fuels start running out? Why?

e In 1990, wind energy cost over 10 p/kWh to generate. How do you think the cost may change in future? Explain your answer.

Scientists have estimated how much it costs to produce 1 kWh of electricity from each energy source. The estimates take account of the different running costs for each power station.

Supplying electricity

Electricity may be needed at any time of the day or night. But some energy sources cannot be relied on to provide a constant supply. We do not know when there will be a sunny day, or the right amount of wind. These forms of energy have low **reliability**.

Question

f Explain in detail why:
(i) solar cells
(ii) wind turbines
(iii) tidal generators
cannot produce a constant electricity supply.

We also have to think about where the electricity is needed. The raw materials must be transported to the power station, and the electricity must be transferred to users.

The Caribbean island of St Lucia makes almost all its electricity from diesel oil. This is a fossil fuel that has to be imported. Electricity costs about twice as much there as in the UK.

Question

g The government of St Lucia plans to generate much more electricity from local wind and solar plants. Do you think this is a good idea? Explain your answer.

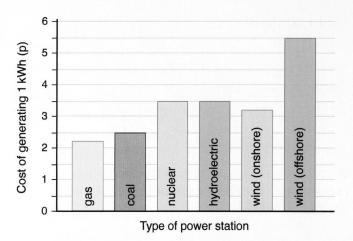

 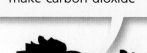 

Gas is cheap

Nuclear power doesn't make carbon dioxide

Wind and waves won't run out

It depends on the situation

Key points

- Energy sources for making electricity have different advantages and disadvantages.
- To choose between them, we have to think about many factors, including the cost of electricity, the reliability of the energy source and the environmental impact.

1 Match the words **A**, **B**, **C** and **D** with the spaces **1–4** in the sentences.

    A particles   **B** conduction
    C thermal energy   **D** convection

There can be no ____**1**____ in a solid, because the ____**2**____ cannot move from place to place. But in some solids ____**3**____ can be transferred efficiently by ____**4**____.

2 A group of students investigated thermal energy conduction in brass, stainless steel, copper and aluminium. The diagram shows how the experiment was set up.

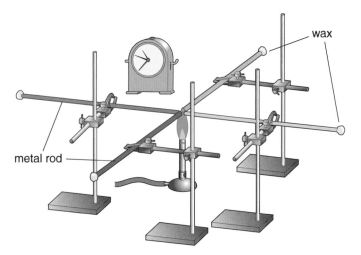

A blob of wax was placed at one end of each rod. The other end of the rod was held in a flame. The students measured how long it took for the wax on each rod to start melting.

The table shows the results.

Metal	Time for wax to start melting (s)
brass	58
stainless steel	103
copper	14
aluminium	29

  a Write down **two** methods of thermal energy transfer apart from conduction. *(2 marks)*

  b Which metal is the best conductor? *(1 mark)*

  c What is the dependent variable in this experiment? *(1 mark)*

  d The rods were all the same size. Explain why this is important. *(3 marks)*

3 The diagram shows a solid cylinder made of metal that has been painted black. The cylinder has been left in the sunshine.

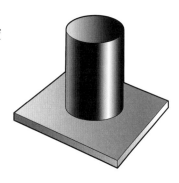

Complete the sentence by using the correct words from the box.

conduction metal radiation emission black solid convection

The cylinder's surface absorbs _____ from the Sun. It can do this well because it is _____. Thermal energy is transferred from the surface to the inside of the cylinder by _____. Energy is also transferred from the surface to the surrounding air, which warms up and rises by _____. *(4 marks)*

4 Some students investigated thermal energy loss from three cubes. The diagram shows how the experiment was set up.

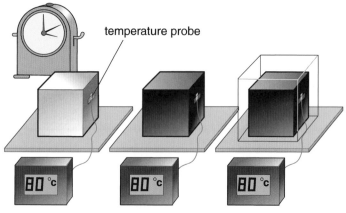

One of the cubes was painted shiny silver. The other two were painted dull black. Each cube had a temperature probe attached to its surface to let the students measure the temperature of the cubes.

The cubes were warmed to 80 °C in an oven, then left out on safety mats to cool down. The silver cube and one of the black cubes were left uncovered. The other black cube was covered with a plastic box.

The students recorded the temperature of each cube every minute as they cooled down. The table shows some of their results.

Silver uncovered		Black uncovered		Black covered	
Time (min)	Temp. (°C)	Time (min)	Temp. (°C)	Time (min)	Temp. (°C)
0	80	0	80	0	80
1	74	1	72	1	75
2	69	2	65	2	70
3	64	3	59	3	66
4	60	4	53	4	62
5	56	5	47	5	58
6	52	6	43	6	55
7	49	7	40	7	52
8	46	8	37	8	49

a The students wanted to investigate the effect of a cube's colour on its rate of cooling. Which two cubes should they compare?
(2 marks)

b State which colour surface is a better emitter of thermal radiation. Explain how you can tell this from the experiment. *(2 marks)*

c Which method of thermal energy transfer is affected by the cover? Explain why. *(2 marks)*

d Using the data in the table, describe the effect of covering a cube on its rate of cooling.
(2 marks)

5 The diagram shows a heat sink from the back of a computer. The heat sink is designed to transfer thermal energy from the computer to the surroundings.

The heat sink is made of metal. The base is in contact with the computer. The fins have a large surface area in contact with the air. The whole computer, including the heat sink, is enclosed in a plastic casing.

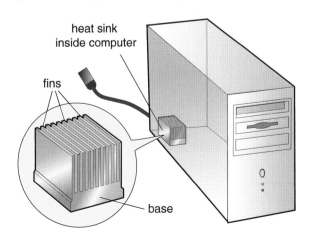

a By which method is thermal energy transferred from the computer to the heat sink base? *(1 mark)*

b The heat sink base is one of the heaviest parts of the computer. A lighter base would make the computer easier to carry. But the manufacturer says that a base with a larger mass transfers thermal energy faster. Is this correct? Explain your answer. *(2 marks)*

c **i** Which method of thermal energy transfer needs air to flow freely around the fins?
(1 mark)

ii Explain why the computer's plastic casing makes the heat sink less good at transferring thermal energy. *(1 mark)*

d The computer has a fan to blow air over the heat sink fins. Explain how this helps the heat sink to transfer thermal energy to the surroundings. *(1 mark)*

6 The diagram shows two species of hare. One lives in hot deserts. The other lives in colder regions where there is snow in the winter.

Which one of the statements **A**, **B**, **C** and **D** is correct?
A The hare with large ears lives in hot regions. Its ears help it to radiate thermal radiation to the surroundings.
B The hare with small ears lives in hot regions. Its ears are compact to allow good convection.
C The hare with large ears lives in cold regions. Its ears help it to absorb thermal radiation from the surroundings.
D The hare with small ears lives in cold regions. Its ears are compact to conduct thermal energy to the rest of the body.

7 Match the words **A**, **B**, **C** and **D** with the definitions 1–4.

    A efficiency   **B** useful energy
    C transfer   **D** power

1 The wanted form of output energy.

2 The movement of energy from one place to another.

3 The proportion of the input energy that is usefully transformed.

4 The rate at which energy is transformed.

8 The pie chart shows the energy output from a heater in an hour.

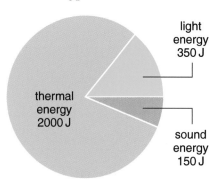

light energy 350 J

thermal energy 2000 J

sound energy 150 J

a Write down the useful form of energy output. *(1 mark)*

b Showing your working, calculate the efficiency of the heater. *(2 marks)*

c What is the total amount of energy supplied to the heater in an hour? Explain how you know. *(2 marks)*

9 The table shows the costs and annual savings of four methods for reducing energy consumption in the home.

	Method	Cost (£)	Annual saving (£)
1	loft insulation	150	50
2	cavity wall insulation	300	100
3	insulation round skirting boards	28	7
4	insulation of solid walls	900	150

a Which of the methods **1**, **2**, **3** and **4** is the most effective at reducing energy consumption?

    A method 1   **B** method 2
    C method 3   **D** method 4

b Which of the methods, **1**, **2**, **3** and **4** are cost-effective within 5 years?

    A 1 and 2   **B** 1, 2 and 3
    C all of them   **D** 4 only

c Which of the methods, **1**, **2**, **3** and **4** would give the greatest overall saving after 5 years?

    A method 1   **B** method 2
    C method 3   **D** method 4

10 The diagram shows some of the energy transformations and transfers that might be carried out in order to heat a room.

Match the processes **A**, **B**, **C** and **D** with the labels **1–4** on the diagram.

    A voltage changed by step-up transformer
    B voltage changed by step-down transformer
    C energy transformed in power station
    D energy transformed in electric heater

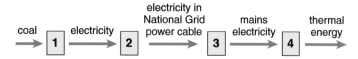

coal → **1** → electricity → **2** → electricity in National Grid power cable → **3** → mains electricity → **4** → thermal energy

11 A 3 kW electric heater is run for 120 minutes each day. The cost of electricity is 6p per unit.

Match the numbers **A**, **B**, **C** and **D** with the sentences **1–4**.

    A 360   **B** 60   **C** 3000   **D** 6

1 The number of units of electricity used each day.

2 The cost in pence of running the heater for 10 days.

3 The power rating of the heater in W.

4 The number of kWh of electricity used in 10 days.

12 The pie chart shows the proportion of the UK's electricity generated from different sources. Which **one** of the sentences **A**, **B**, **C** and **D** is correct, according to the data in the chart?

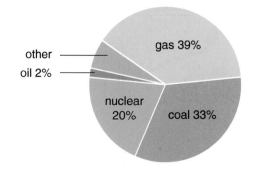

other

oil 2%

gas 39%

nuclear 20%

coal 33%

A 6% of the electricity came from hydroelectric plants.

B At least 94% of the electricity came from fossil fuels.

C No more than 6% of the electricity came from renewables.

D None of the electricity came from solar power.

13 The diagram shows some parts of a hydroelectric power station.

Match the labels **A**, **B**, **C** and **D** with the descriptions **1–4**.

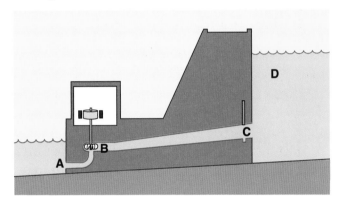

1 The place where the dam can be opened or closed.

2 The place where the water has the most gravitational potential energy.

3 The place where the water turns a turbine.

4 The place where the water leaves the hydroelectric plant.

14 The table shows how much carbon dioxide is produced when electricity is generated from various sources.

Energy source	Carbon dioxide emission (g/kWh)
gas	500
coal	900
nuclear	25

a Write down **one** way in which carbon dioxide affects the environment. *(1 mark)*

b Which energy source produces least carbon dioxide? *(1 mark)*

c Explain the reason for this. *(1 mark)*

d Explain **two** disadvantages of the energy source you gave in part **b**. *(2 marks)*

15 Turbogen Electrical Company wants to build a power station in the countryside near Anytown. Some local people are in favour of the plan, but others are against it.

The passage below is part of a newspaper article discussing the plan.

> The proposal has split the community. The National Association for Bird Protection is opposed to the plan. 'Birds will fly into the turbines and be killed', said the head of the local branch.
>
> But some environmental groups are in favour. The local council recently dropped plans for a coal-fired power station. 'That would have been a disaster', said an environmental protestor. 'The new plans are a big improvement.'
>
> Meanwhile, the company that wants to build the power station denies that many birds will be killed. The head of research at Turbogen says that the turbines will have less impact on birds than a coal-fired power station. 'Coal pollution is bad for birds', she said. 'Our power station will release no harmful gases'.

Use information from the passage to help you to answer these questions.

a What type of power station is the article discussing? *(1 mark)*

b Give **two** advantages of this energy source for generating electricity. *(2 marks)*

c Suggest **one** reason why environmental groups might prefer this type of power station to a coal-fired station. *(1 mark)*

d Write down **one** issue that the National Association for Bird Protection and the electricity company disagree about. *(1 mark)*

e Suggest how they could investigate this issue to find out the facts. *(2 marks)*

Waves carry energy from one place to another. Most waves are invisible, yet we are constantly surrounded by them. We use them every day of our lives, usually without realising it, and life would be impossible without them.

▲ An infra red photograph of Earth.

The photograph above is not the usual one that you might expect to see from space – it was taken using infra red. This sort of technology helps us to see things that we would not normally be able to see.

We all know that we get light energy from the Sun. Light energy travels to us as a wave and we 'see' it with our eyes. There are all sorts of other waves, similar to light, that we cannot 'see' in the same way – but we can put them to good use provided we can control the dangers they present. They can help us to communicate and to find out what things are like inside, where we can't see.

Is there a tenth planet?

In July 2005 a team of scientists claimed to have discovered a tenth planet. Several times in the previous few years a similar claim had been made but was later found out not to be true. How do scientists make and then check such discoveries? The answer is that they use a range of different telescopes that use the whole of the electromagnetic spectrum – from the shortest gamma rays to the longest radio waves. Each type of radiation can give us different information, and each needs to be carefully checked. In August 2005, some scientists began to have doubts about the latest claim to finding a tenth planet.

TENTH PLANET DISCOVERED!

July 2005

Astronomers have discovered another planet in our solar system. It is bigger than Pluto, but about three times further away from the Sun. It was discovered using telescopes.

IS THERE REALLY A TENTH PLANET?

August 2005

As well as detecting light reflected off the tenth planet, scientists are also using infra red and microwaves to confirm whether last month's discovery really is another planet. At the moment, the infra red telescopes have been unable to detect any radiation, but this could be because a mistake had been made in pointing them in the right direction. Until these checks have been made it cannot be confirmed

Think about what you will find out in this section

What are electromagnetic waves?	How can electromagnetic waves be used?
What are the dangers of electromagnetic waves?	Are mobile phones dangerous to use?
Is it dangerous to sunbathe?	Is it dangerous to live close to a TV transmitter?
Why shouldn't I have too many X-rays taken?	

Watch out – waves about!

In December 2004, an earthquake below the Indian Ocean caused a tsunami to hit the coastlines of southeast Asia. This devastating wave carried enough energy to destroy villages and kill thousands of people.

▲ A tsunami can travel at speeds of up to 600 mph.

If you are unlucky enough to be close to a tsunami, you can see and hear the wave and even feel the energy it carries. But there are some waves that are harder to detect, even though they are all around us. These waves are caused by changes in an electric field and they are called electromagnetic waves.

Tuning in

When you select channels on your radio or TV, you are choosing the wavelength or the frequency of the radio signal you want to receive. Your radio will pick up only the waves that are exactly the frequency you have chosen. But what do we mean by frequency and wavelength? You can picture these more easily by looking at a wave in water.

> #### Question
>
> **a** Where did the tsunami's energy come from originally?

▲ Radio waves belong to the family of waves called the electromagnetic spectrum.

If the waves are 3 m apart, we say the wavelength is 3 m. The number of waves hitting the surfboard every second is the frequency. For example, if one wave hits the surfboard every second, we say the frequency is 1 per second or 1 hertz.

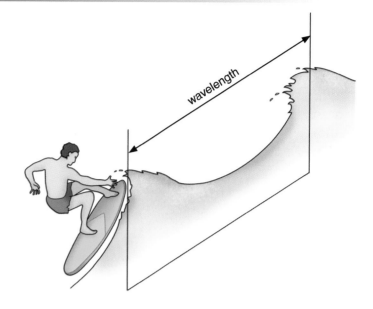

Questions

b What is the frequency if three waves hit the surfboard every second?

c A radio station broadcasts waves with a frequency of 20 kilohertz (kHz). How many waves are produced each second? (Hint: Remember 'kilo' means '1000'.)

Same but different

You can think of the waves in the electromagnetic **spectrum** as members of a family because they resemble each other in some ways, but are different in others. For example, they can all travel through a vacuum (empty space) and they travel through a vacuum at the same speed. The speed of radio waves, infra red, ultra violet and X-rays is the same as the speed of light – 186 000 miles per second or 300 million metres per second.

The family of waves forms a continuous spectrum, just as the colours in a rainbow are continuous. They behave differently because their wavelengths and frequencies are different. The wavelengths vary from several kilometres (radio waves) to tiny fractions of a millimetre (e.g. light and X-rays).

▶ We see the different wavelengths of light as different colours.

Visible light is a section of the electromagnetic spectrum.

Question

d Suggest two ways of detecting visible light.

Key points

- Waves carry energy.
- Electromagnetic waves are caused by disturbances in an electric field.
- All electromagnetic waves travel at the same speed in a vacuum.
- Electromagnetic waves have different wavelengths and frequencies.

The wave formula

We have seen that the frequency of a wave means how many of these cycles pass a given point every second. A **frequency** of one cycle per second is called one hertz (Hz).

The **wavelength** is the distance from any point on the **cycle** to the same point on the next cycle.

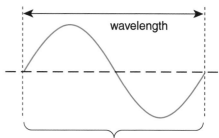

This is one cycle of a wave and its length is called the wavelength. The whole is made up of many of these, repeated over and over again.

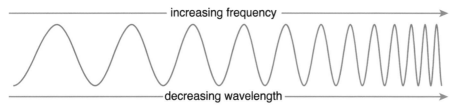

increasing frequency

decreasing wavelength

▲ As the frequency of a wave changes, so does the wavelength.

You can see from the diagram above that as the frequency gets higher, the wavelength gets smaller. These two measurements are related by the formula:

wave speed = frequency × wavelength
(metres/second)   (hertz)   (metres)

Example
A wave has a frequency of 2 Hz and a wavelength of 5 m. What is its speed?

Use the formula:

wave speed = frequency × wavelength

= 2 Hz × 5 m/s

= 10 m/s

▲ A microwave oven.

Questions

a Radio 4 broadcasts on the long wave band. It has a frequency of 200 000 Hz and a wavelength of 1500 m. What is the speed of the waves?

b A microwave oven produces waves of wavelength 0.12 m and frequency 2500 MHz. Do these values support the view that microwaves and radio waves travel at the same speed? (Hint: MHz (megahertz) means 1 000 000 Hz.)

Which wave?

We have seen that waves in the **electromagnetic spectrum** have different wavelengths and frequencies, but how does this make them different? They are produced in different ways, they have different properties and because of their properties they have different uses and dangers.

gamma rays	X-rays	ultra violet	visible light	infra red	microwaves	radio waves

— increasing wavelength ——————→
decreasing frequency

Radio waves have the longest wavelength, ranging from about a metre to several kilometres. Approximately 1 million infra red waves could fit into one radio wave, and approximately 1 million gamma waves could fit into one infra red wave.

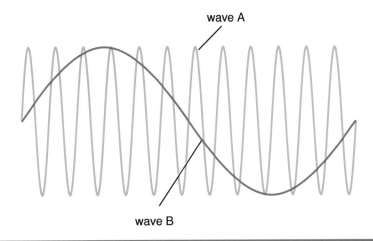

wave A

wave B

Question

c Which wave has: (i) the higher frequency (ii) the longer wavelength?

Generally, the higher the frequency of a wave, the more energy it carries and the more dangerous it is likely to be.

Questions

d Which two types of electromagnetic wave are likely to be most dangerous?
e Some cancer treatments use wave energy to kill cancer cells. Which type of wave might be used for this?
f In the 1930s, workmen fixing a powerful radio transmitter noticed that a hot dog they had left close to the transmitter became hot. Which type of wave do you think was responsible for heating up the hot dog?
g What does this tell you about low frequency electromagnetic waves?

Key points

- All electromagnetic waves travel at the same speed in a vacuum (300 million m/s).
- Electromagnetic waves obey the wave formula:
  wave speed = frequency × wavelength
- The uses and dangers of electromagnetic waves depend on their wavelengths and frequencies.

Electromagnetic waves carry energy through empty space (a vacuum), but what happens to all this energy when the wave hits a material?

In practice, three things can happen to waves: **reflection**, **absorption** or **transmission**.

Reflection

This means that the wave bounces off in a different direction.

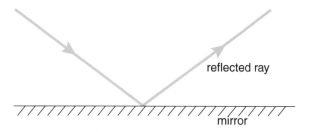

reflected ray

mirror

▲ Light can be reflected from a mirror.

▲ Light waves can be reflected from smooth surfaces such as the calm water of a lake.

> **Question**
>
> **a** Suggest some other surfaces that are good at reflecting light. What do these surfaces have in common?

Absorption

Thick curtains can stop heat radiation (infra red) escaping from a room at night. They do this by absorbing the infra red. When a wave is absorbed, the energy goes into warming up the material and does not get through.

Water molecules are good at absorbing microwave energy. When you cook food in a microwave oven, the energy from the microwaves makes the water molecules vibrate more vigorously, so raising the temperature. The heat generated in the surface of the food (the top 3 cm) is then conducted through to the inside.

If you want to prove that it is the water that is heated, try drying an ice cube and putting in the microwave oven. Switch on, and you will find that the ice does not melt!

Radio receiver aerials absorb the energy from radio waves. The radio waves produce a tiny alternating current in the aerial. The radio receiver turns this into the sound that you hear from the speaker.

▲ Bodies can sometimes get in the way of radio waves!

> **Question**
>
> **b** Where do the microwaves get their energy from?

Transmission

Radio waves can travel through cloud, light waves can travel through glass and water, and some X-rays can get through soft tissue but not bone. Different materials transmit different electromagnetic waves.

For example, glass is excellent at transmitting light and that is why it is used in windows.

Some materials, like glass, will reflect, absorb and transmit electromagnetic radiation. Glass absorbs very little light, but it is quite good at absorbing other wavelengths such as ultra violet (so you won't get a suntan sitting indoors behind the window!) It can also be quite good at reflecting light, as you may have noticed when you see your reflection in a shop window. You can still see what is behind the shop window because much of the light is transmitted.

▲ A sheet of glass can make a good reflector.

Question

c Look at the Thermoglass leaflet. What percentage of heat (infra red) radiation will be transmitted into the office?

Key points

- Waves can be reflected, absorbed or transmitted.
- Different materials are good at reflecting, absorbing or transmitting different wavelengths.
- When radiation is absorbed, the energy makes the substance hotter, and may create an alternating electric current of the same frequency as the radiation itself.

▲ You can still receive satellite TV broadcasts on a cloudy day!

THERMOGLASS

- This glass is ideal for modern office blocks

5%

80%

- It absorbs 80% of the heat (infra red) radiation and reflects 5%

- This stops most of the heat being transmitted into your office on a hot summer day

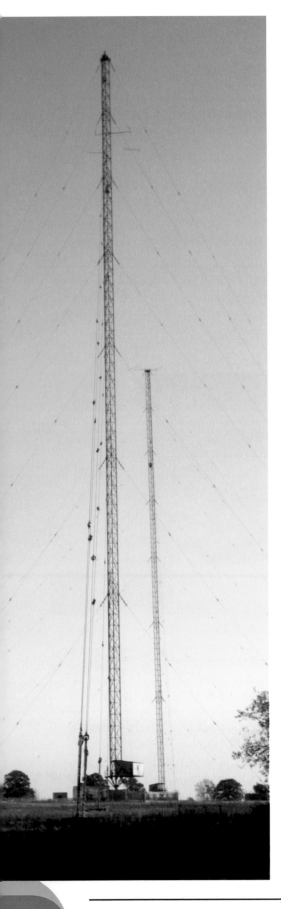

Radio waves

◀ These aerials at Droitwich are half as high as the wavelengths being transmitted.

Radio waves have the longest wavelength of all the electromagnetic waves. They include waves for television as well as radio.

Microwaves

Microwaves are useful for communications because they can pass easily through the atmosphere.

▲ The sensor at the top of this sign uses microwaves to monitor traffic speed. They send messages to car drivers, warning them if they are speeding.

▲ Radar is used on ships and at airports.

Long wavelength microwaves are also used for radar.

Microwaves are sent out from the dish. When they hit an object such as an aeroplane, they bounce back. The time taken between sending out the waves and detecting their return is used to work out the position of the object.

Question

a *What property of microwaves makes them suitable for radar?*

Microwaves and mobile phones

Here is a news item about mobile phones.

ARE MOBILE PHONES SAFE?

There is no evidence of ill effects caused by radiation at current national safety limits, but gaps in current knowledge mean mobile phones cannot be classed as 'safe' yet.

Microwaves may affect body cells but this may not always lead to disease or injury.

Children should be especially careful about mobile phone use. Their skulls are still growing, and their cells are also developing and tend to absorb more radiation compared to adults. This is especially important in brain tissue. Children should only use mobiles for short, essential calls.

How do we communicate?

Before modern communication systems, Morse code was used to send light or radio signals. It consists of a series of short bursts (dots) and long bursts (dashes). Each letter of the alphabet is represented by a different series of dots and dashes. Here is a very important signal – SOS.

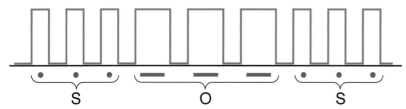

Digital and analogue signals

A **digital signal** is one that makes sudden jumps between on and off. An **analogue signal** can vary continuously between off and a maximum. For example, you could send an analogue message by turning a dimmer switch up and down to vary the brightness of a lamp.

Question

c *What could you do to a lamp to send a digital signal?*

The first mobile phones, radio and television used analogue signals. Analogue signals are more likely to be affected by interference so they are being replaced by digital signals. These give higher quality and carry more information. Sometimes signals have to be changed from digital to analogue or vice versa. Modern mobile phones change your analogue voice signal into a digital signal before transmission.

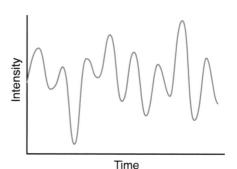

▲ An analogue wave.

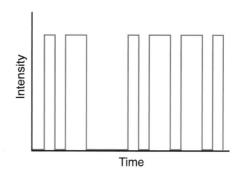

▲ A digital signal.

Key points

- Microwaves and radio are widely used in communications.
- Communication signals can be analogue or digital.
- Digital signals tend to be of higher quality and can be processed by computers.

Visible light

Photocells are often used when small amounts of electricity are needed and there is no mains supply.

▶ This parking meter gets its power from photocells that convert **visible light** into electricity.

Ultra violet waves

▲ The ultra violet lights used by crime scene investigators can reveal hidden blood, fingerprints fibres, and bruises on living and dead bodies.

Certain substances fluoresce (or glow) under **ultra violet** light. For example, some washing powders contain an ingredient that makes clothes look 'whiter-than-white' when seen in sunlight. You can use a special pen to write your postcode on your valuable possessions – you can't see it in normal light, but it shows up under ultra violet.

Many of us like to be outside in the sunshine – but be careful! Too much exposure to the ultra violet light in sunlight can cause damage to living cells, leading to skin cancer.

Try to stay out of the sun between 11am and 3pm in the summer.

Apply sunscreen of factor 15 or greater 15–30 minutes before going outside and reapply it every 2 hours

Question

a Here is part of a leaflet giving advice about sunbathing.
(i) Why is it more important to stay out of the sun at the times mentioned?
(ii) Why do you think that sunscreen needs to be reapplied after every 2 hours?

Infra red radiation

This is sometimes called heat radiation because hot objects emit more infra red radiation than cold objects. **Infra red** sensors can detect heat from the body, and so are used for burglar alarms and night-time photography. Infra red is also used in remote controls for televisions and cars.

◀ The photo on the right is taken with a thermal imaging camera using infra red light; the photo on the left is of the same building using visible light.

Question

b Why do you think that some of the areas in the right-hand picture are much brighter and whiter than others?

Fibre optics

optical fibre

ray of light

▲ The light reflects off the inside surface of the fibre.

Modern technology uses light or infra red to send waves along optical fibres made of glass. These fibre optics can channel the light over long distances and round corners with very little loss of energy.

Doctors use fibre optics when they use an endoscope for looking inside a person's stomach. A fibre optic cable is carefully passed down the patient's throat into the stomach. The fibre optic can carry light down into the stomach, where it is reflected. It then travels back up the fibre optic to form an image.

◀ Optical fibres allow light to travel along curved paths.

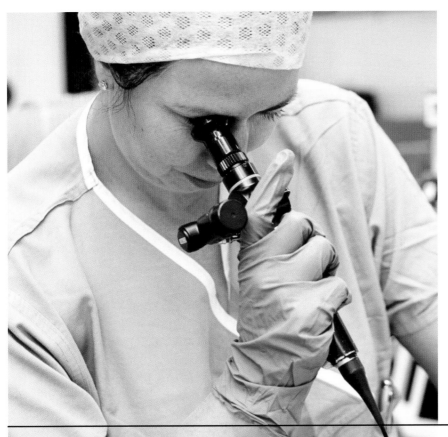

Key points

- Ultra violet rays can be harmful to living cells.
- Ultra violet light can cause fluorescence.
- Visible and infra red light can be used to send signals along optical fibres.

X-rays

Doctors often use **X-rays** to help in diagnosing different conditions. X-rays can penetrate soft tissue quite easily, but are absorbed by more dense materials such as bone and metal. They are also used at airport security checkpoints to look inside luggage.

X-rays generally have a longer wavelength than gamma rays. As with all electromagnetic waves, they can be dangerous. Over-exposure to X-rays can cause cell damage. This is why radiographers stand behind a protective lead shield when they take an X-ray (lead is good at absorbing X-rays). They may also wear badges that check how much radiation they have been exposed to.

> **Question**
>
> **a** In the 1950s most shoe shops had X-ray machines for checking that children's shoes fitted properly. The child, the shop assistant and the parent could all look into the machine to see the child's foot inside the shoe. Why do you think that these machines are no longer used?

▲ These X-ray machines were common in shoe shops in the 1950s.

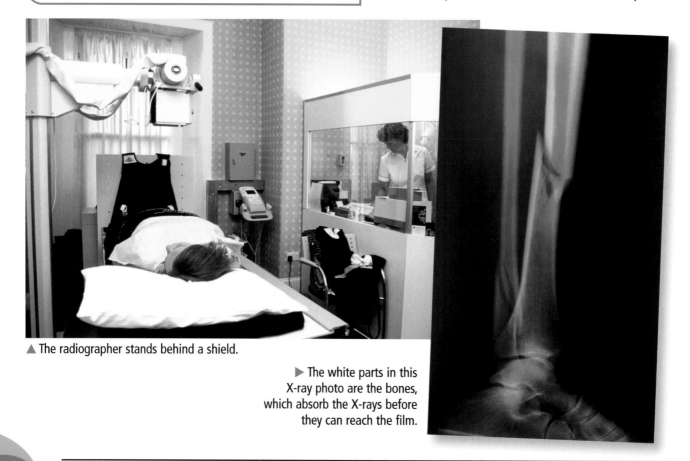

▲ The radiographer stands behind a shield.

▶ The white parts in this X-ray photo are the bones, which absorb the X-rays before they can reach the film.

Gamma rays

Gamma rays have the shortest wavelength and the highest frequency, and carry the most energy. They are produced by some radioactive substances and by stars.

Because of the high amount of energy they carry, they can kill cells. This makes them very dangerous but the effect can be used to kill cancer cells.

▶ Gamma rays can be used to detect faults in a pipeline.

Gamma radiation can also be used for checking for cracks in cast metals or checking that welds in steel pipes have been made correctly. Because of its high energy, gamma radiation is very penetrating. This means that it can pass through thick metals that other types of radiation cannot. When the radiation meets a crack, the waves change speed and are partially reflected, and these changes can be detected.

Question

b Gamma rays have a very short wavelength. What does this tell you about the frequency of gamma rays?

Gamma rays can be used to irradiate food to stop it from going bad. The gamma rays kill bacteria that cause food to decay quickly. Some people believe that irradiating food produces chemicals that harm us. Others say that although gamma rays may kill the bacteria on the food, they do not destroy the toxins that these bacteria have already produced, and it is these toxins that make us ill.

Ingredients: A blend of Thyme; Sage; Origanum and Marjoram (irradiated).

Ideal to complement the flavour of pizza, tomato, meat, chicken and fish dishes.

Recipe Suggestion: BUTTERMILK HERB BREAD: Preheat oven to 180 °C. Sift together 500 g Cake Flour, 5 ml Salt, 25 ml Baking Powder, and 10 ml Choice Mixed Herbs. Beat together 1 Egg, 50 ml Oil and 500 ml Buttermilk. Make a well in the centre of the dry ingredients and pour in the buttermilk mixture. Mix well to form a fairly soft dough. Place in a greased loaf tin and bake for approximately 50 minutes.

Storage Instructions: Keep away from direct sunlight. Remove foil sachet, clip corner and refill your Choice Mixed Herbs bottle.

PRODUCT OF SOUTH AFRICA

▲ Food that has been irradiated should carry a label saying that this has been done.

Question

c Why is it important that food that has been irradiated should be labelled like this?

Key points

- The hazards of electromagnetic waves depend on the type of wave and the dose received.
- Gamma rays and X-rays have a high frequency and a short wavelength.
- Because gamma rays and X-rays are very penetrating, they are used for diagnostic purposes and for killing cancer cells and bacteria.
- People who work with gamma rays and X-rays need to take special precautions.

Microwaves

Here is part of the instruction booklet from a microwave oven.

Using your microwave oven safely

 Inspect the door seals regularly to make sure they are not worn.

 Walk away from the oven when it is in use – do not stand very close to it.

Question

a Suggest reasons for each piece of advice shown.

Radio waves

Some people are worried about living next to powerful radio and television transmitters, just as they worry about mobile phone masts.

▶ A powerful TV transmitter.

NO INCREASED LEUKAEMIA RISK FROM TRANSMITTERS

People living close to radio or television transmitting masts are not at increased risk of developing leukaemia, say the results of a national study published recently.

Question

b There is still a lot of debate about whether powerful radio and television transmitters may be a danger to health.
(i) What sort of evidence do you think that scientists would need in order to make a statement like the one above?
(ii) If the statement had been made by the broadcasting station rather than by scientists, would this make any difference to whether you believed it? Explain your answer.

Very low frequency waves

Whenever an alternating current goes through a wire it produces an alternating electromagnetic field. This in turn produces electromagnetic waves.

The frequency of the alternating current mains supply is 50 Hz. This means that near to overhead power lines there will be electromagnetic waves of frequency 50 Hz.

> **Question**
>
> **c** Use the wave formula to calculate the wavelength of electromagnetic waves whose frequency is 50 Hz. In which part of the electromagnetic spectrum do these waves belong?

At present, no one is absolutely sure if these waves have any effect on our bodies. However, some people who live very close to National Grid lines are worried that there may be some long-term effects.

> **Question**
>
> **d** Here are two different newspaper articles about the effects that living very close to overhead power lines might have.
> (i) Explain why Mrs Jones is wrong to say that 'this proves that power lines cause cancer'.
> (ii) Does the research quoted in article A prove that power lines are safe? Explain your answer.

No danger from overhead power lines

Recent studies in Australia have found no evidence of a link between cancer and power lines. No increases in the incidence of cancer were detected in mice after prolonged exposure to the type of fields emitted by power lines.

▲ Article A.

Second death in family sparks new fears over power lines

John and Jane Jones have lost their second son to cancer this week. Jack (18) died almost a year to the day since his younger brother Jim died of the same disease. 'This proves that these power lines are causing cancer,' Mrs Jones told our reporter.

▲ Article B.

> **Key points**
>
> ● It is possible to collect data that can be analysed in order to provide evidence for scientific theories.
> ● It is necessary to distinguish between opinion based on valid and reliable evidence and opinion based on non-scientific ideas.

Everything is made up of atoms, and all atoms have a nucleus. So why are people worried when the word 'nuclear' is mentioned?

▲ What does this mean? There are no nuclei in this town?

Radiation! Nuclear! Atomic!

Some people are frightened of the very words 'radiation', 'atomic' and 'nuclear'. They think of atomic bombs and the terrible effects of nuclear explosions. Yet we and all the things around us are made up of atoms and nuclei. Nuclear reactions are happening all the time in the Sun – without them we would not receive any warmth or light.

Background radiation

Radioactivity occurs naturally all the time. We are all exposed to low levels of radiation every day. This is called **background radiation**. Some rocks, such as granite, are radioactive. If you live in Aberdeen, you are exposed to much more radiation than if you live in London. Naturally occurring uranium in the rocks in Aberdeen releases a radioactive gas called radon. If radon gas collects under the floor of your house, you may have a slightly higher risk of developing cancer.

▲ Should we all have one of these on our front door?

Cosmic radiation

A lot of background radiation comes from space and is called **cosmic radiation**. Most of it gets filtered out by the atmosphere, but some of it gets through. Airline pilots and cabin crew spend a lot of time at 30 000 feet where the atmosphere is thinner, and so are exposed to a lot more cosmic radiation.

Everyday sources of radiation

Some background radiation comes from man-made sources, such as nuclear power stations, X-ray machines and luminous watches. Most of us have a smoke alarm fitted in our homes. The most common type contains a weak radioactive source.

The pie chart below shows the percentage of background radiation that comes from different sources.

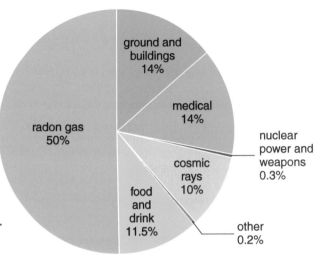

▶ You receive four or five times as much radiation from natural sources as you do from man-made sources.

Risks and dangers

Scientists have learned how to control some nuclear reactions. Nuclear reactions can be very useful for generating electricity but very destructive if used to make bombs. The waste products made by these reactions can give off radiation for years, leading to increased risks of cancer. It is important that we understand about radioactivity so that we know how to deal with it.

Think about what you will find out in this section

What are atoms made from?	Why do some atoms split up into pieces?
What are the different kinds of radioactivity?	What is meant by half-life?
How can we use radioactivity?	What are the dangers of radioactivity?
How can we protect ourselves from radioactivity?	

What a load of rubbish!

Most human activity produces waste, but the waste from nuclear reactors can cause a lot more problems than ordinary household rubbish.

British radioactive waste stockpile still growing

Britain has over 1000m³ of high-level nuclear waste. This is enough to completely fill 20 average school science labs. For lower-level waste, there is 300 times as much. The waste is stored at more than 20 different sites around the country, and some of it will be radioactive for more than 200000 years. As more nuclear power stations are shut down and taken out of service, this amount will rise – perhaps by as much as 50 times over the next 10 or 20 years. This represents a potential danger to the public and the Government is being urged to come up with a plan for dealing with this.

Questions

a Give two reasons why nuclear waste may be more dangerous than any other kind of waste material.
b Look at the pie chart on the previous page. What percentage of the background radiation comes from medical sources?

Rates of decay

Some nuclear waste contains a radioactive substance called plutonium. The plutonium atom is unstable – this means that it splits up to make a new atom and at the same time gives out some radioactivity. We call this process **radioactive decay**.

The decay takes place at different rates, depending on the substance. Some substances take thousands of years to decay completely, but others take only fractions of a second.

The decay is unpredictable – we can never know for sure when a particular atom will decay, but we can estimate when *half* the atoms in a particular sample will have decayed. This is called the **half-life**.

Question

c Do you think that plutonium takes a very long time or a very short time to decay? Give a reason for your answer, using information from the newspaper cutting above.

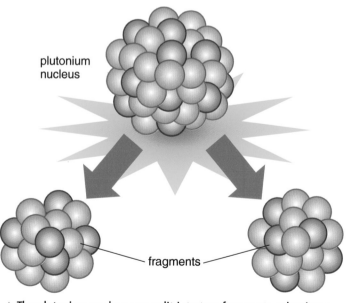

plutonium nucleus

fragments

▲ The plutonium nucleus can split into two fragments, releasing a lot of energy at the same time.

The half-life is the time taken for the **count rate** to fall to half of its current value. The count rate is a measure of how radioactive a substance is. Count rate can be measured by a Geiger counter, a device that gives a click or records a number every time it detects some radiation.

The graph shows how the rate of decay changes with time.

You can see on the graph how the half-life has been calculated.

At the start of the timing, the count rate was 1000 counts per minute. Half of 1000 is 500. Draw a line from 500 counts per minute across to the graph line. Then draw a line down to the time axis. It meets it at 5 minutes, and so the half-life is 5 minutes.

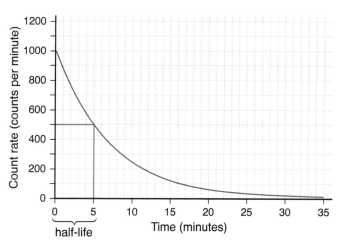

▲ How half-life is calculated.

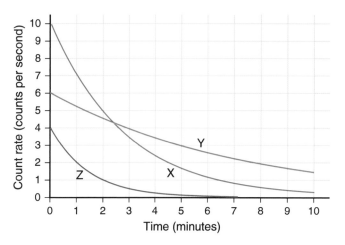

This second graph shows the rate of decay for three different substances.

Questions

d Which substance has the longest half-life?
e What is the half-life of substance X?

Remember that it doesn't matter how many atoms are present to start with – for any one substance, it always takes the same length of time for half of them to decay.

Atoms decay spontaneously – no matter what happens to them. Things such as temperature, pressure or chemical reactions do not affect the rate of decay.

Key points

- Some nuclei are unstable and give out radiation. These substances are said to be radioactive.
- The rate of radioactive decay does not depend on pressure or temperature.
- The half-life of a substance is the time it takes for half the atoms in a sample to decay.
- A half-life may be fractions of a second or thousands of years, depending on the substance.

Radioactivity in the home

▲ A typical smoke detector.

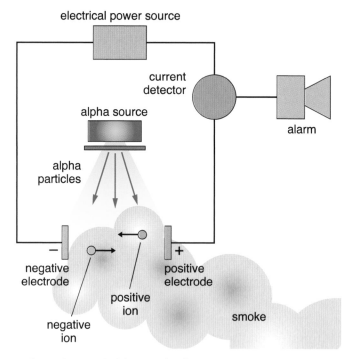

▲ The main parts inside a smoke detector.

Americium-241 is a radioactive metal. It gives off a type of radioactivity called alpha particles. A tiny amount of the metal is placed in a smoke detector. A 9 V battery is also needed to power a monitoring circuit. The alpha particles bombard the molecules in the air and help them to create a tiny electric current. If there is a fire, smoke particles get inside the detector and stop the electric current. As soon as the current falls, the monitoring circuit turns on an alarm.

Looking inside the atom

Americium-241 is one type of americium atom. The number 241 tells us exactly what is in the nucleus, or central bit, of the atom.

The **nucleus** of an americium-241 atom contains 95 **protons** and 146 **neutrons** – that is 241 particles altogether. A different type of americium atom is called americium-243. These different types are called **isotopes**.

> **Question**
>
> **a** All americium atoms have 95 protons in the nucleus. How many neutrons does americium-243 have?

The nucleus of americium-241 is unstable. This means it will split up to make a new nucleus and at the same time give out some radioactivity.

> **Question**
>
> **b** Alpha particles are absorbed by only a few centimetres of air, but they are very damaging to living cells. Use this information to explain why an alpha source is safe to have in a smoke detector, but why you should never take a smoke detector apart.

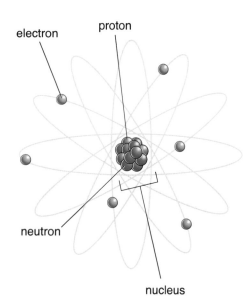

▲ An atom is made up of protons, neutrons and electrons.

Three types of radioactivity

Americium-241 gives out **alpha particles**, but there are two other types of radioactive decay: **beta particles** and **gamma rays**. Different substances are chosen for different uses depending on the type of radioactivity they emit.

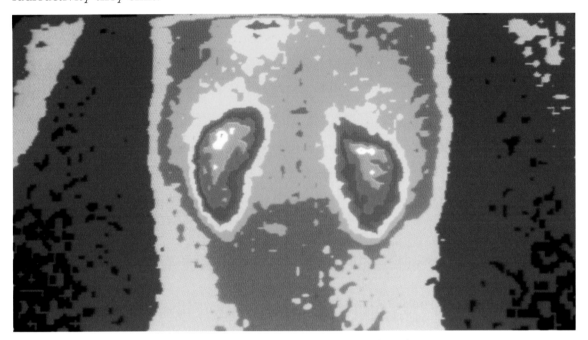

▲ This photo shows a scan of a two healthy kidneys. The centre of the kidneys (orange/white) have absorbed the tracer, technetium-99m.

Some radioactive substances can be used as **tracers** in medicine. The substance is injected into a patient and the radiation it gives off can be tracked as it moves around the body. For example, a blocked kidney can be shown up by injecting a solution of radioactive technetium-99m into a patient. The kidney takes up the technetium, and this causes it to show up on film sensitive to the radiation. A doctor uses the photograph to help in deciding how to treat the patient.

Substances used as tracers in medicine should have short half-lives and emit gamma radiation.

Questions

c All technetium atoms have 43 protons in the nucleus. How many neutrons are in the nucleus of the technetium-99 isotope?

d Some technetium atoms have only 54 neutrons. Choose the name of this type of technetium from the list below.

    A   technetium-43
    B   technetium-54
    C   technetium-97
    D   technetium-153

e Why is it important that the isotope used as a tracer should have a short half-life?

Key points

- All atoms are made up of a small central nucleus surrounded by electrons.
- The nucleus of an atom is made up of neutrons and protons.

Alpha, beta or gamma?

There are three types of radioactivity: alpha, beta and gamma, named after the first three letters of the Greek alphabet. But how do scientists decide which type to use for each application?

The three different types of radiation have different properties.

- Alpha particles are absorbed in the air within a few centimetres and can be stopped completely by a sheet of thin paper.
- Beta particles are more penetrating but can be stopped by thick cardboard or a thin metal sheet.
- Gamma rays can travel easily through concrete walls or several centimetres of lead shield.

Radioactivity can be used to control the thickness of aluminium foil produced by machines.

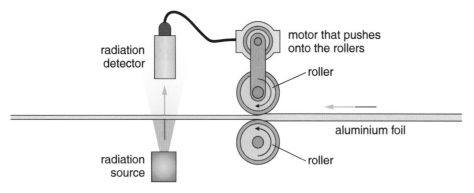

Radiation from the source passes through the foil to the detector. The thicker the foil, the less radiation gets through. The thinner the foil, the more radiation gets through. The detector senses this and sends signals to the motors to put more or less pressure on the rollers. If the foil is too thick, the rollers will push harder on the foil to make it thinner.

Question

a Use information about the three types of radiation to explain which source you would choose for this machine: alpha, beta or gamma.

Ionisation

Although not very penetrating, alpha particles cause a large amount of **ionisation**. This is the process of making ions. Ions are atoms that have lost or gained **electrons** and so are electrically charged. Ionising radiation is very dangerous to living tissue. Alpha particles are the same as helium nuclei (two protons and two neutrons). They can be made to change direction by using electric or magnetic fields.

Beta radiation consists of electrons that have come from the nucleus of a decaying atom. They too are affected by electric and magnetic fields, but do not cause as much ionisation as alpha particles.

Gamma radiation is not affected by electric and magnetic fields, and causes the least amount of ionisation. Gamma rays are a form of electromagnetic radiation.

Gamma radiation can be used to sterilise medical equipment. Items such as artificial hip joints, surgical gowns or dressings are placed in sealed packets and passed beneath the source so that they are exposed to radiation. The radiation sterilises the materials by killing the bacteria.

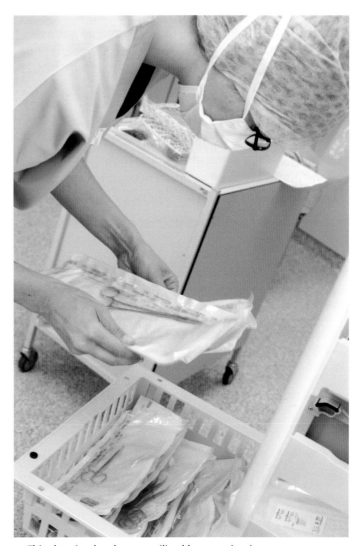

▲ This dressing has been sterilised by exposing it to gamma rays.

Questions

b Why is it important that the gamma source for this type of application should have a long half-life?

c Irradiation is also used to sterilise food and cosmetics. Why do you think this is a better method than sterilising by boiling?

d Imagine that you are a vet investigating the health problem of an expensive racehorse. You need to inject a radioactive tracer and can choose between polonium-210 (alpha) and cobalt-60 (gamma). Explain which you would choose and why.

Key points

- The three types of radiation are alpha particles, beta particles and gamma rays.
- Alpha particles are very ionising, but not very penetrating. Alpha particles are the same as helium nuclei.
- Beta particles produce less ionisation, but are more penetrating than alpha particles. Beta particles are electrons that have been produced by a decaying nucleus.
- Gamma rays are very penetrating. They can only be stopped by several centimetres of lead or several metres of concrete. Gamma rays are a form of electromagnetic radiation.

What are the dangers?

All radioactivity is potentially dangerous, but how dangerous it is depends on:

- the type of radiation (alpha, beta or gamma)
- the intensity (strength) of the radiation.

Gamma radiation

At low intensities, gamma radiation can damage cells, but at high intensities it kills cells. Gamma radiation can be used to kill cancer cells by focusing a beam of gamma rays onto them.

▶ The intensity is greatest at about 5 mm below the surface. The normal cells receive a much lower dose.

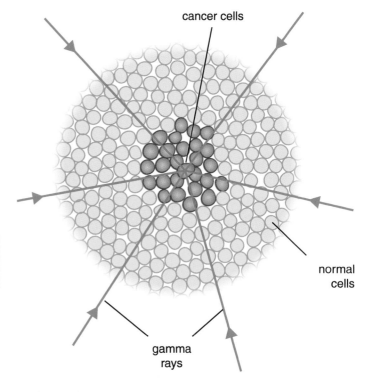

cancer cells

normal cells

gamma rays

Question

a Why does the cancer cell receive a much higher dose than the normal cells?

X-rays

X-rays are very helpful in showing doctors what is going on inside our bodies, but they have to be used with care. Too much exposure could increase our risk of cancer; but what are the risks?

Type of X-ray	Equivalent period of background radiation	Extra lifetime risk of cancer in addition to the 1 in 3 chance we all have of getting cancer
chest or teeth X-ray	a few days	less than 1 in 1 000 000
head X-ray	a few weeks	between 1 in 1 000 000 and 1 in 100 000
breast (mammography)	a few months to a year	between 1 in 100 000 and 1 in 10 000
stomach (barium meal)	a few years	between 1 in 10 000 and 1 in 1 000

To put the risks into perspective, a 4-hour plane flight carries about the same extra risk as a chest X-ray.

Questions

b What is the chance that we all risk of getting cancer, even without any extra exposure to radiation?

c Which type of X-ray carries the greatest extra risk?

d Which of the four types of X-ray, shown in the table, would pregnant women not normally be given?

Monitoring the dose

People who work with radioactive materials need to be protected from possible harm and monitored to make sure that they have not been exposed to too much radiation.

▶ This type of badge is worn by people who work with radioactive materials.

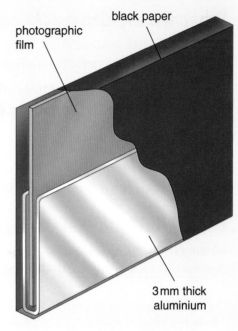

▲ The more radiation that reaches the film, the darker the film becomes.

Questions

e *Name one industry or profession where a worker might need to wear one of these badges.*

f *How does this badge work? Choose the correct answer from the list below:*
- *It absorbs all the radiation before it can reach the body.*
- *It reflects all the radiation away from the body.*
- *It warns if too much radiation has been received.*

Dealing with waste

The more that we use nuclear fuels and radioactive materials, the more radioactive waste builds up. Some of this can remain a danger for thousands of years and we need to decide how to dispose of the waste safely.

Here are some suggestions.

Suggested method	Advantage
Put it on a spaceship and fire it into the Sun.	Gets it away from the Earth altogether.
Put it into thick metal tanks.	Removes it well away from where people are living.
Bury it at the bottom of a disused coalmine.	
Vitrification – that is turning it into a glass-like material.	Stops the material from being washed into the water supply by rain.

Nobody has yet come up with the perfect solution. What has to be done is to carefully consider the advantages and disadvantages of using such materials, and then try to make a rational decision based on the best scientific evidence available.

Questions

g *Suggest an advantage of burying the waste at the bottom of a disused coalmine.*

h *For each of the methods in the table, suggest one disadvantage.*

Key points

- It is important to evaluate the benefits, drawbacks and risks of scientific and technological developments.
- Decisions about radiation hazards are likely to have social, economic and environmental effects, and may raise ethical issues.

For centuries we have been gazing at the stars and pondering the mysteries of the Universe. Many different theories have been put forward to describe and explain how it all works. It is only since the beginning of space travel that we have gathered enough data to be able to suggest many of the answers.

How did our Universe begin?

In the twentieth century there were two main theories put forward to explain the origin of the Universe. One, produced by Professor Fred Hoyle, was called the steady state theory. This said that although matter seemed to be disappearing, it was continually being replaced.

Another theory was the 'big bang' theory. This stated that all the matter in the Universe was originally at one point. A tremendous explosion occurred and ever since the material has been travelling outwards.

Scientists continue to collect evidence in order to decide which is the better theory.

▲ The Crab Nebula is the remnant of a huge explosion called a supernova. Although this happened over 6000 years ago, it was not seen on Earth until 1054. At that time it was bright enough to be seen in the daytime, as bright as 400 million Suns.

▼ The Earth from space.

Making observations

Modern scientific instruments have allowed us to work out the structure of our solar system with its nine planets revolving around the Sun. We now know that our Sun is just one of millions of stars that make up our galaxy (the Milky Way), and that there are millions of such galaxies in the Universe.

When you look at the stars, you are actually looking back in time. Even at the incredibly fast speed that light travels, it takes about 8 minutes to reach us from the Sun, and about 4 years to reach us from the next nearest star. Light from some of the stars has taken many millions of years to reach us.

But the big questions are – how did it all start and how is it changing now?

Scientists think that the Universe is about 15 billion (15 000 000 000) years old. By studying the light and other types of radiation from distant galaxies, we can work out what the Universe must have been like at the very beginning. It may have started as a primordial 'soup' of incredibly dense material and energy.

Special telescopes, such as the Hubble telescope, have been launched into space to collect information that would be impossible to collect on Earth.

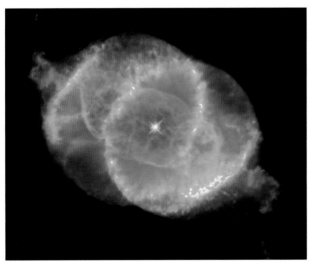

◄ Photos taken from space are clearer – the radiation does not have to travel through the dust and gases in the Earth's atmosphere.

Think about what you will find out in this section

How do scientists use telescopes of different sorts to observe the Universe?	How can electromagnetic waves be used to study the Universe?
What is the red-shift and why is it important?	What sort of radiation do stars give out?
Is the Universe expanding?	How do we know the age of the Universe?

The first telescopes

Ancient peoples observed the skies using their eyes only. They were able to predict the movements of comets and other celestial bodies, but there was a limit to the amount they could see, until the development of telescopes in the last few hundred years. The first telescopes used the visible part of the electromagnetic spectrum – in other words, light.

In the seventeenth century, Galileo used a refracting telescope to observe four of Jupiter's moons. This enabled him to confirm an earlier theory of Copernicus which stated that the Earth and other planets revolved around the Sun.

A refracting telescope uses lenses to focus light into an image.

One of the problems with lenses is that they absorb some of the light. Also, unless the lenses are shaped perfectly, they can make the image distorted. This prompted Isaac Newton, a few years after Galileo, to develop a new kind of telescope called a reflecting telescope. This used a concave mirror in addition to a lens to focus light into an image.

> **Question**
>
> **a** Give one reason why a telescope that uses a mirror may be better than one that only uses lenses.

Modern telescopes

The problem with all telescopes sited on the Earth is that the atmosphere tends to get in the way and reduces the amount of radiation reaching the telescope. That is why in 1990 NASA launched the Hubble telescope. This telescope was mounted on a satellite orbiting the Earth above the atmosphere.

Optical (light) telescopes on Earth cannot be used during the day, as there is too much light from the Sun. Cloud cover can also affect their images.

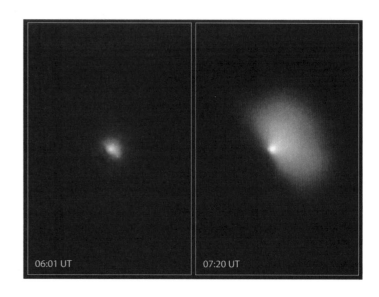

▶ The Deep Impact spacecraft fired a massive projectile straight into a comet to find out what it was made of. This picture of the impact was taken with the Hubble telescope.

06:01 UT   07:20 UT

Not all telescopes use visible light to investigate the Universe. This is because stars (including the Sun) give off radiation throughout the entire electromagnetic spectrum. Hubble, for example, can detect ultra violet and infra red radiation, as well as visible.

NASA has launched the Chandra X-ray observatory on a satellite – it is particularly good at detecting the existence of 'black holes'.

Question

b *Suggest one advantage and one disadvantage of mounting a telescope on a satellite rather than on the Earth.*

▲ This radio telescope at Jodrell Bank can be used at any time of the day or night.

Key points

- Telescopes may detect visible light or other electromagnetic waves such as infra red, radio or gamma radiation.
- Observations of the solar system, the Milky Way and other galaxies can be carried out on Earth or from space.

The red-shift

Astronomers discovered that the Universe appears to be expanding. They found this out by looking at the frequencies of the radiation from distant galaxies.

Have you ever noticed how, when an ambulance or police car has its siren blaring, the pitch of the note seems to be higher when it is coming towards you and lower when it is going away from you?

This effect applies to all sorts of waves, sound waves as well as electromagnetic waves.

Remember that the frequency of a wave means how many cycles reach you every second. With sound waves, the higher the frequency, the higher the pitch.

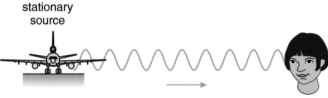

stationary
source

▲ If you are standing still in front of a source of waves, then the frequency with which the waves reach you is constant. If it's a sound, the pitch stays the same; if it's a light source, the colour stays the same.

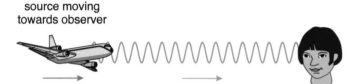

source moving
towards observer

▲ If the source of the waves is coming towards you, then the cycles reach you sooner than they would otherwise. In other words, the frequency appears to have increased. This means that a sound wave would have a higher pitch or a light source would appear to be more blue.

source moving
away from observer

▲ If the source of the waves is moving away from you, then the waves reach you later. In other words, the frequency appears to have decreased. This means that a sound wave would have a lower pitch or a light source would appear to be more red.

You can easily notice the effect with sound, but you do not normally notice the effect with light. This is because light waves have such a small wavelength that the source of light has to be moving very fast for it to be noticeable. Astronomers have noticed an effect similar to this with light from stars. It is called the **red-shift**.

Light tells a story

The first diagram shows the spectrum of light from the Sun – our closest star. Various elements in the Sun cause the black lines in it.

The second diagram shows the spectrum for a much more distant star. Notice how the black lines have been shifted to the right (red end) of the spectrum.

Astronomers have concluded that the stars are moving away from us. They have also found out that the further away the stars are from us, the faster they seem to be moving.

Scientists can get a lot of information from studying the spectrum of a star. Not only can they work out how far away it is, but they can also tell which chemical elements it contains.

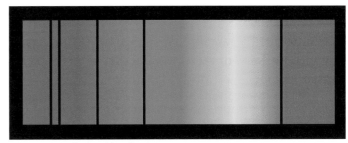

▲ Spectrum formed by the Sun.

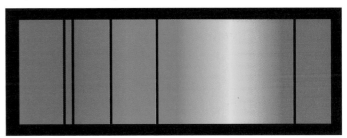

▲ Spectrum formed from a distant star.

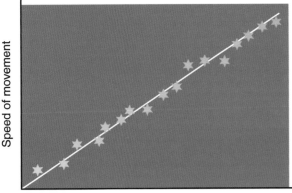

Distance from Earth

Name of star	Distance from Earth in light years
Alpha Centauri	4.27
Deneb	1600
Sirius	8.64

Questions

a The table above shows the names of three stars and their distances from us.

Which star would you expect to show:
(i) the smallest red-shift
(ii) the largest red-shift?
Explain your choices.

b Look at the graph on the right, which shows the speed at which stars are moving away from us, plotted against distance from us. What does this graph tell you about how these two quantities are related?

Key points

- If a wave source is moving, a stationary observer will observe a change in wavelength and frequency. This effect causes a red-shift in light from distant galaxies.
- The further away a galaxy is, the bigger the red-shift.

Is the Universe changing?

In the twentieth century, astronomers became occupied with questions about the origins of the Universe. Two main theories emerged that dealt with how the Universe might be changing – the steady state theory and the big bang theory.

The big bang

Einstein's theory of relativity had predicted that the Universe should be expanding, and in 1912 an American scientist called Vesto Slipher noticed a red-shift in the spectra of light from many galaxies. A red-shift would mean that these galaxies are moving away from us. Later, Edwin Hubble was able to compare the amount of red-shift with the distance he had measured to some of these galaxies, and was able to show that the further away a galaxy was, the faster it appeared to be moving. This led other scientists to develop the **big bang theory**.

▼ Edwin Hubble.

If the Universe is expanding, there must have been some time at which all the matter in the Universe was compressed into a single point (some scientists reckon about the size of a pea). This incredibly dense and incredibly small lump of matter exploded, and the Universe has been expanding ever since. By measuring how far away a galaxy is, and knowing the speed it is moving, we can work out how long it has been moving. According to this theory, the Universe is about 14 billion years old.

Question

a *What two pieces of information do we need to work out how long a galaxy has been moving for?*

▼ Stars are forming in the red and blue glowing gases of the Lagoon Nebula.

The steady state theory

Other scientists did not like the idea that the Universe had been created at some point in time – they believed that it must have always existed. Sir Fred Hoyle developed the **steady state theory**, which states that as the Universe expands and the galaxies get further apart, new galaxies are being formed. This would keep the density of the Universe constant, or in a steady state.

Which theory is correct?

A good scientific theory should be able to make predictions. We can then look for some evidence to see if these predictions are correct.

- The big bang theory states that the Universe was once tiny, with a lot of energy packed into it. As the matter and energy spread out, the temperature would have dropped and we should be able to detect the radiation from the early beginnings of the Universe. In 1965 this was detected and is now known as cosmic microwave background radiation.

- The other prediction of the big bang theory is that in the first couple of minutes after the big bang, conditions would be just right to produce hydrogen and helium in the ratio of about 4 to 1. This is exactly what we find today when we look at the spectra of the most distant stars. Light from these stars has been travelling for the longest time and tells us what conditions were like when these stars were first made.

Question

b Suggest two pieces of evidence that support the big bang theory.

What created the Universe in the first place?

This is one of those questions that science will probably never be able to answer.

▲ One of the early models of our solar system.

Key points

- Uncertainties in scientific knowledge and ideas change over time.
- There are some questions that science cannot currently answer and some that science cannot address.

1 This question is about the use of X-rays on hospital patients.

Match words **A, B, C** and **D** with the spaces **1–4** in the sentences.

    A bone   **B** cells
    C flesh   **D** lead

X-rays are used because they are absorbed by ____**1**____ but travel straight through ____**2**____. The radiographer stands behind a shield made of ____**3**____ because constant exposure to X-rays is dangerous and could damage ____**4**____ in the body.

2 The diagram shows the electromagnetic spectrum.

Radio waves	Microwaves	Infra red	Light	Ultra violet	X-rays	Gamma rays
	1	**2**		**3**		**4**

Match the uses, **A, B, C** and **D** to the numbers in the diagram.

    A Checking welds on pipes
    B Detecting forged banknotes
    C Operating mobile phones
    D Television remote controls

3 This question is about electromagnetic waves.

Match words **A, B, C** and **D** with the spaces **1–4** in the paragraph.

    A energy   **B** frequency
    C speed   **D** wavelength

All electromagnetic waves carry ____**1**____ and travel at the same ____**2**____. Those with the shortest ____**3**____ are called gamma rays and those with the lowest ____**4**____ are called radio waves.

4 The diagram shows Jack shaking the end of a rope to make waves.

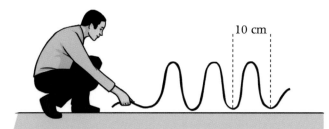

10 cm

a How many complete waves (cycles) are shown in the diagram? *(1 mark)*

b Use the scale on the diagram to measure the wavelength of this wave. *(1 mark)*

c Jack shakes the rope at a frequency of 2 hertz. Use the wave formula to work out the speed of the wave. *(1 mark)*

5 A factory has been investigating the best area of solar cells to use for a remote traffic signal. They want the maximum voltage output, but the cheapest cost. The panels cost £10 per 100 cm². Some of the results are shown below.

Surface area exposed (cm²)	Average voltage produced (mV)
0	0
20	40
40	80
60	100
80	115
100	120
120	125
140	130
160	133
180	134
200	135

a Which part of the electromagnetic spectrum does a solar cell use? *(1 mark)*

b Which measurement in the table was
    i the independent variable *(1 mark)*
    ii the dependent variable? *(1 mark)*

c Draw a graph of the results shown in the table. *(2 marks)*

d Describe in detail the way in which the average voltage produced changes with the amount of surface area exposed. *(1 mark)*

e How much would it cost to use
    i a 100 cm² solar panel for the traffic signal *(1 mark)*
    ii a 200 cm² panel for the traffic signal? *(1 mark)*

f The company producing the panels recommends that the traffic signal should use as large an area of panel as possible. Do you think that this recommendation is fair? *(1 mark)*

6 A glass manufacturer produces four different types of glass for use in windows.

Glass type	Percentage of infra red radiation from the Sun that is:		
	Absorbed	Reflected	Transmitted
Sologlass	60	5	?
Heliglass	70	10	20
Superglass	80	?	5
Wonderglass	?	8	7

a What percentage of infra red (IR) radiation does Sologlass transmit? *(1 mark)*

b What percentage of IR radiation does Superglass reflect? *(1 mark)*

c What percentage of IR radiation does Wonderglass absorb? *(1 mark)*

d What happens to the IR radiation that is absorbed by the glass? *(1 mark)*

e Name one other type of radiation that reaches us from the Sun. *(1 mark)*

7 Sunscreen creams are given an SPF number – SPF stands for Sun Protection Factor. The number tells you how many times longer you can stay out in the sun compared with if you did not put any cream on.

Without sunscreen you could stay out for 10 minutes without burning.

a How long could you stay out for if you used a 5 SPF sunscreen? *(1 mark)*

b Other than becoming sunburned, what other damage could be caused to your skin by over-exposure? *(1 mark)*

c What type of electromagnetic radiation are sunscreens designed to protect you from? *(1 mark)*

d What might happen to your bathing towel if it were exposed to this radiation? *(1 mark)*

e The instructions say that if you are sitting near water or on a light sandy beach you should reduce your exposure times. What is the reason for this? *(1 mark)*

8 The table shows the half-lives of four different radioactive forms of iodine.

Radioactive iodine is sometimes injected into the human body to help diagnose disease.

Form	Symbol	Half-life
P	I-108	36 milliseconds
Q	I-117	2.2 minutes
R	I-129	15 million years
S	I-135	6.5 hours

a What is the correct name for these different forms of iodine? *(1 mark)*

b What is the meaning of the term half-life? *(1 mark)*

c How many milliseconds are there in a second? *(1 mark)*

d Starting with 1 g of I-108, how long would it be before only 0.25 g was left? *(1 mark)*

e Starting with 1 g of I-129, how much would be left after 30 million years? *(1 mark)*

f **i** Which form would it be best to use in the human body? *(1 mark)*

    ii Explain the reason for your choice. *(1 mark)*

9 The diagram represents an atom.

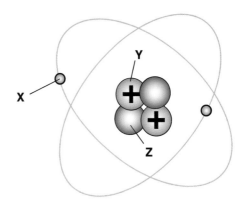

a Name the particles labelled
    i X *(1 mark)*
    ii Y *(1 mark)*
    iii Z *(1 mark)*

b What name is given to the central part of the atom? *(1 mark)*

10 The diagram shows three types of radiation, labelled **P**, **Q** and **R**.

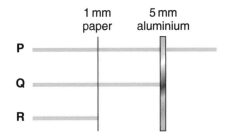

a What type of radiation is
i P, **ii** Q and **iii** R? *(3 marks)*

b Which type of radiation could have its direction of travel changed by
i a magnetic field *(2 marks)*
ii an electric field? *(2 marks)*

11

The Gloworm Bathroom Heater

Warm your bathroom by infra red radiation. Central heating radiators first warm the air and then the warm air heats you! Our infra red Gloworm heaters transfer the heat directly to you! The heat is instant!

Mrs Jones sees the advert and says, 'I don't like the sound of radiation – I heard that thousands died of radiation after that disaster at the Chernobyl nuclear power station. I think I'll stick to my water-filled central heating.'

a One of the nuclear reactors at the Chernobyl nuclear power station exploded in 1986. Immediately people were evacuated from within a 20 mile radius.
i What kind of radiation caused the deaths of people after the Chernobyl disaster? *(1 mark)*

ii Why do you think that they did not evacuate people who lived more than 20 miles away? *(1 mark)*
iii Why do you think that the site at Chernobyl is still dangerous after 20 years? *(1 mark)*

b What could you say to Mrs Jones to reassure her that the Gloworm Heater was safe? *(2 marks)*

12 We receive X-rays all the time from natural sources.

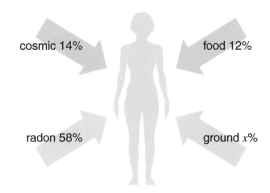

▲ Most of the X-rays we receive come from natural sources.

a Which of the four sources shown in the diagram gives us the largest amount of X-rays? *(1 mark)*
b What percentage of the X-rays we receive come from the ground? *(1 mark)*
c Where do cosmic rays come from? *(1 mark)*
d Radon is a radioactive gas that emits alpha particles. Normally it is only present in extremely tiny amounts in the atmosphere, and causes us no problem. In large amounts it can be very dangerous.
Why does the fact that it is **i** a gas and **ii** an alpha emitter make it especially dangerous? *(2 marks)*

13 The diagram below shows one kind of optical telescope.

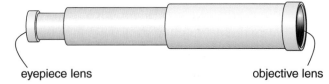

eyepiece lens   objective lens

a What kind of optical telescope is this? *(1 mark)*
b What kind of electromagnetic radiation does it use? *(1 mark)*
c This kind of telescope can be used on Earth to study the stars.
i Give one advantage of using this type of telescope compared with any other type. *(1 mark)*
ii Give one disadvantage of using this type of telescope compared with any other type. *(1 mark)*

d **i** Name **one** other type of telescope that could be used to study the stars. *(1 mark)*

ii Give **one** advantage that the telescope that you have named has compared with the one shown in the diagram. *(1 mark)*

e Some telescopes are mounted on satellites in space rather than on the surface of the Earth. What is the advantage of this? *(1 mark)*

14 The radiation given out by some stars appears to have been shifted towards one end of the spectrum.

a Which end of the spectrum does the radiation appear to have been shifted towards? *(1 mark)*

b What does this tell you about the position of the star? *(1 mark)*

c Why do astronomers find this effect particularly interesting? *(1 mark)*

15 For many years scientists have been trying to understand the origin of the Universe, and what is happening to it at present.

a What are the names of the **two** main theories about the Universe? *(2 marks)*

b Which of these two theories is currently thought to be the better explanation? *(1 mark)*

c What was the name of the scientist who proposed the steady state theory? *(1 mark)*

d Give **one** piece of evidence that supports the big bang theory. *(1 mark)*

16 The table shows the half-life of some radioactive isotopes.

Radioactive isotope	Half-life
lawrencium-257	8 seconds
sodium-24	15 hours
sulphur-35	87 days
carbon-14	5600 years
plutonium-239	24360 years
chlorine-36	300000 years
uranium-235	731000000
potassium-40	1300000000 years

a Starting with the same amount of each, which of the isotopes in the table would take the longest time to decay?

b What is the meaning of the word isotope? *(1 mark)*

c What does the number after the isotopes name tell you, e.g. sodium-24? *(1 mark)*

d How long would it take for 100 grams of lawrencium-257 to decay to the stage where there was only 50 grams left? *(1 mark)*

e You start with 10 grams of chlorine-36. How much would be left after 300000 years? *(1 mark)*

f **i** Which of the isotopes in the table could be used as a tracer in medicine? *(1 mark)*

ii Explain the reason for your choice. *(1 mark)*

g Carbon-14 can be used to date ancient skeletons.

This is done by measuring the amount of carbon-14 in the ancient skeleton and comparing it with the amount in a living skeleton.

Why would this method be of no use to a forensic scientist who wanted to find out when a person had died during the last 5 years? *(2 marks)*

absorb To take in radiation.

absorption When waves lose energy while travelling through a material.

acid rain Rain that has been acidified by pollutants such as sulfur dioxide.

adaptations A feature or features that make a structure more suitable for its function.

addicted Dependant on a drug.

alkanes Hydrocarbons that only have single C-C or C-H bonds, e.g. ethane C_2H_6.

alkenes Hydrocarbons with one (or more) C=C double bonds, e.g. ethane C_2H_4.

alloy Carefully blended metal mixture with specific properties.

alpha particle Two protons and two neutrons – the same as a helium nucleus.

analogue signal A signal that has a shape corresponding to the shape of the signal being carried; the opposite of a digital signal.

antibiotics Chemicals produced by microbes, used to destroy bacteria in the body.

antibodies Substances produced when white blood cells detect the presence of a particular antigen.

antitoxins Substances produced by white blood cells to neutralise poisons produced by microbes.

arthritis A painful inflammation of the joints.

asexual reproduction Reproduction that does not involve the formation of gametes. New organisms formed by asexual reproduction are genetically identical to the parent organisms.

atmosphere The layer of gas that surrounds the Earth; approximately 80% nitrogen, 20% oxygen.

atoms The smallest parts of an element that still have the properties of that element.

background radiation Radiation that we are exposed to all the time, from radioactive substances around us, and from cosmic rays.

bacteria A type of microbe. Some bacteria are useful, others cause disease.

beta particle An electron emitted from the nucleus of an atom during radioactive decay.

big bang theory The theory that says the Universe was thought to have been created by a huge explosion.

biodegradable Something that can be broken down easily by natural, biological processes.

biodiesel A renewable fuel made from vegetable oil that can be used in place of diesel.

carbon monoxide Poisonous gas produced by incomplete combustion of carbon compounds.

carcinogens Substances that cause cancer.

cement A powder that sets hard when mixed with water; made by heating limestone with clay.

central nervous system The brain and spinal cord.

characteristics Distinguishing features.

chemical bonds Bonds that form between atoms during chemical reactions, and which hold molecules and other compounds together.

chromatography A method of separating different dyes and chemicals such as food colours.

chromosomes Long threads containing many genes, found in the nucleus of a cell.

clinical trial An investigation to provide data on the effectiveness of a drug.

clones Genetically identical organisms.

concrete A building material made by mixing cement, sand and gravel with water: sets to a very hard 'artificial rock'.

conduction Transfer of thermal energy by transfer of vibration from particles to their neighbours.

conductors Substances that are good at conducting thermal energy.

consumers Organisms that feed upon those below them in a food chain.

continental drift The exceedingly slow movement of the continents across the surface of the Earth.

contraceptive pill A pill containing hormones that prevent an egg being released.

convection Transfer of thermal energy in a gas or liquid by the movement of particles from place to place. Hotter, less dense regions float and cooler, denser regions sink.

convection currents Swirling currents that form when a liquid or gas (or very hot rock) is heated from below.

cosmic radiation Radiation from the early beginnings of the Universe.

cost-effective A measure is cost-effective if it saves more money than it costs.

count rate The reading on a Geiger counter showing the rate at which radiation has been detected.

cracking The process by which long-chain hydrocarbons are broken up into shorter and more useful hydrocarbons.

crust The thin, hard and brittle outer layer of the Earth.

cycle A series of processes that always occur in the same order.

deficiency disease A disease caused by the lack of a substance in the diet.

deforestation Clearing of trees.

diabetes Disorder where the pancreas fails to control glucose concentration.

digital signal A signal that carries information in the form of a string of on and off pulses; the opposite of an analogue signal.

effective A measure is effective if it has a great effect. An energy-saving measure is more effective the more money it saves.

effector An organ or cell that brings about a response to a stimulus.

efficiency The proportion of the energy supplied to a device that is transformed usefully rather than wasted.

efficient The higher the efficiency of a device, the more efficient it is.

egg The female sex cell (gamete).

electrolysis The tearing apart of a molten (or dissolved) ionic compound using electricity.

electromagnetic spectrum All types of electromagnetic radiation, arranged in order according to their wavelengths and frequencies.

electromagnetic wave A form of energy that is transferred as fast-moving waves. Electromagnetic waves include light and thermal radiation and can travel through a vacuum.

electrons Small particles with tiny mass and a single negative charge.

embryo transplants Transferring embryos from one organism and implanting them into another uterus (womb).

embryos Unborn offspring.

emit To give out radiation.

emulsifiers Chemicals that help stop an emulsion from separating.

emulsion A mixture of tiny droplets of oil in water (or water in oil).

enzymes Catalysts produced by cells.

ethanol The 'alcohol' in alcoholic drinks made by fermenting sugar; may also be used as a fuel.

extinct A species that used to live on earth but which has all died out.

fertility drug A drug used to increase the chances of a woman becoming pregnant.

flu epidemic Outbreak of a highly contagious viral infection of the respiratory passages, spreading to many people.

fossil fuels Fuels such as coal, oil or gas, formed over millions of years from the remains of living things.

fractional distillation A form of distillation where only a partial separation of the liquids in a mixture is obtained.

fractions The different liquids produced from a complex mixture such as crude oil by fractional distillation.

frequency The number of waves per second; measured in hertz (Hz).

FSH Follicle-stimulating hormone; stimulates eggs to mature and the production of oestrogen.

gametes Specialised sex cells involved in sexual reproduction in plants and animals.

gamma rays Part of the electromagnetic spectrum of waves that have the shortest wavelength.

gels Materials with a grid-like structure that can trap water.

generator A device that produces electricity when it spins. Power stations contain generators.

genes Parts of a chromosome that control an inherited characteristic.

genetic modification The deliberate modification of the characteristics of an organism by manipulating its genetic material.

geothermal Using steam heated by hot underground rocks to make electricity.

global dimming Sunlight reaching the Earth is weakened as a result of atmospheric pollution.

global warming The gradual increase in the overall temperature of the Earth's atmosphere caused by increasing levels of greenhouse gases such as carbon dioxide and methane.

GM crops Genetically modified crops. These are crop plants that have had new genes added from another species.

greenhouse gases Gases such as carbon dioxide that help to trap heat energy in the atmosphere.

half-life The average time taken for half the atoms in a sample of a radioactive substance to decay.

herbicides Chemicals used to kill weeds.

hormones Chemicals that are transported around the body in the blood. These chemicals control body processes.

hydrocarbons Compounds made of carbon and hydrogen atoms only.

hydroelectric Using water flowing down a hill to make electricity.

hydrogels Strong gels that can be used to make contact lenses.

hydrogenation A chemical reaction where a hydrogen molecule joins with another compound, e.g. the hydrogenation of unsaturated oil to make saturated fat.

immune Protected against disease by the production of antibodies.

immunised Given a vaccine, containing dead or inactive pathogen, which stimulates the immune system to produce antibodies and memory cells.

impulses Form in which information is transmitted by nerve cells.

infra red A type of electromagnetic radiation that transfers thermal energy.

infra red radiation Energy spreading out from a hot object in the form of waves.

ingesting Taking food into the body.

inherited Characteristic transmitted from parents to children via gametes (eggs and sperm).

insulators Substances that are poor at conducting thermal energy.

ionisation An ionised atom or molecule is electrically charged because it has lost or gained electrons.

ions Charged particles.

isotopes Atoms of an element come in different forms, depending on the numbers of neutrons they have in their nuclei.

Kyoto treaty A treaty established in 1997, under the United Nations, which requires its signatories to reduce emissions of all greenhouse gases.

lattice The regular arrangement of particles in a crystal, for example.

LH Luteinising hormone, which stimulates egg release.

lipoprotein Proteins that are combined with fats or other lipids.

mantle The middle, soft, rocky layer of the Earth that can move very slowly.

mass number The total number of protons and neutrons in the nucleus of an atom.

menstrual cycle The monthly cycle of changes in a woman's reproductive system, controlled by hormones.

metabolic rate A measure of the energy used by an animal in a given time period.

microwaves Part of the electromagnetic spectrum of waves that have wavelengths from about 10 cm to 0.1 mm.

monomers Small molecules that are joined up to form a polymer.

mortar A paste made from slaked lime and water that was once used to stick bricks together.

National Grid The system of power stations, cables and transformers that transfer electricity all over the country.

nerves Bundles of nerve cells.

neurones Cells specialised to transmit electrical nerve impulses and so carry information from one part of the body to another.

neutrons Sub-atomic particles with no electric charge and a relative mass of 1.

noble gases Gases from Group 0 of the Periodic table, e.g. helium, argon, neon.

non-renewable Resources that, once used, cannot be replaced.

nuclear fission The splitting of radioactive atoms. Nuclear fission releases energy.

nucleus The central part of the atom containing the proton(s) and, for all except hydrogen, the neutrons; has most of the mass of the atom.

ores Natural compounds of a metal from which the metal can be extracted.

painkillers Drugs that reduce pain.

pandemic A disease occurring over a wide area.

pathogens Microorganisms that cause disease in plants or animals.

penicillin Antibiotic produced by the mould Penicillium.

pesticides Chemicals that kill pests such as insects.

plate tectonics The theory used to explain continental drift.

polymerisation The chemical process that joins monomers together to make a polymer.

polymers Very long-chained molecules made by joining lots of small molecules together.

power The rate at which energy is transformed.

power stations Factories for producing electricity.

protons Sub-atomic particles with a positive charge and a relative mass of 1.

quicklime Calcium oxide (CaO), a strong alkali formed by heating limestone.

radiation Energy spreading out from a source (e.g. infra red or light energy), or carried by particles (e.g. from a radioactive substance).

radio waves Part of the electromagnetic spectrum of waves that have the longest wavelength.

radioactive A radioactive material contains some atoms whose nuclei are unstable, and may spontaneously break down giving out radiation.

radioactive decay When a radioactive atom decays, it emits radiation and becomes a different type of atom.

receptors Organs or cells that are sensitive to external stimuli.

recycle To re-use materials over and over again.

red-shift The change in wavelength of light from a distant star; it looks redder because it is receding.

reduction The removal of oxygen from a compound, e.g. iron oxide is reduced to iron.

reflection When waves bounce off a surface.

reflex action Rapid involuntary response to a particular stimulus.

reliability The proportion of the time that a device is working. Some types of power station have greater reliability than others because they can work more of the time.

renewable An energy source that will not run out.

resistance Ability to not be destroyed by the action of antibiotics.

sampling The process of taking a small proportion of a larger population as being representative of the population as a whole.

saturated A carbon-chain molecule that only has single bonds between the carbon atoms.

sense organs Organs of the body that are sensitive to external stimuli.

sexual reproduction Biological process of reproduction involving the combination of genetic material from two parents.

slaked lime Calcium hydroxide ($Ca(OH)_2$), made by adding water to quicklime.

solar electric cell A device that converts light directly into electricity.

spectrum A series of waves, arranged in order according to their wavelengths and frequencies.

sperm The male sex cell (gamete).

statins Drugs that act to reduce levels of cholesterol in the blood.

steady state theory A theory of the Universe that states that matter is continually being replaced as it spreads out.

step-down transformer A device that changes a high voltage into a lower one.

step-up transformer A device that changes a low voltage into a higher one.

stimulus A detectable change.

sustainable development Development that conserves natural resources.

symptoms Effects of a disease on the body, such as high temperature and a runny nose.

synapse The junction between two nerve cells.

tectonic plates Massive sections of the Earth's crust that gradually move around the Earth's surface, transporting the continents.

temperature difference The difference in temperature between one object and another.

Thalidomide Drug formerly used as a sedative, but found to cause abnormalities in the developing fetus.

theory of natural selection The theory that evolution occurs by the organisms best adapted to the environment surviving to breed and pass on their genes to the next generation.

thermal decomposition Breaking down a chemical compound by heating.

thermal energy A form of energy; heat. The hotter an object, the more thermal energy it has.

tissue culture A cloning technique that involves growing groups of cells into new plants.

tracers A radioactive isotope that is used in medicine to follow the course of a biological process (e.g. digestion), providing information about the events in the process.

transects Lines or strips used to sample a habitat where environmental conditions change, for example the seashore.

transferred Movement of energy from place to place.

transform To change energy from one form to another.

transformers Devices that change the voltage of an electricity supply.

transition metals The 'everyday' metals such as iron and copper, found in the central block of the Periodic table.

transmission When waves are able to pass through a material.

tsunami A large wave generated by an undersea earthquake.

turbine A device that is designed to turn in order to spin a generator.

ultra violet Part of the electromagnetic spectrum of waves that have wavelengths that are just shorter than those of visible light.

unsaturated A carbon-chain molecule that has at least one double bond between its carbon atoms.

useful energy The energy output from a device that is in the form we want.

vaccine Preparation made from dead or inactive pathogens, which can be injected so that the body makes antibodies to destroy live pathogens of that type.

viruses Types of microbes that cause disease; examples of such diseases are measles and the common cold.

visible light A very narrow band in the middle of the electromagnetic spectrum that enables us to see.

wasted energy The energy output from a device that is in a form we do not want.

wavelength The length of a single wave, measured from one wave crest to the next.

white blood cells Blood cells that counteract infection.

withdrawal symptoms Symptoms in a drug-dependent person who stops taking the drug or reduces the dosage.

X-rays Part of the electromagnetic spectrum of waves that have very short wavelengths, similar to the diameter of atoms.

Index

Revision Guides

Beat the rest - exam success with Heinemann

Ideal for homework and revision exercises, these differentiated **Revision Guides** contain everything needed for exam success.

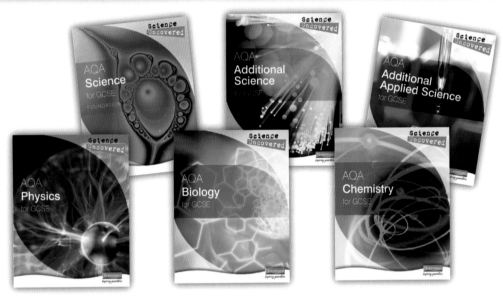

✓ Personalised learning activities enable students to review what they have learnt

✓ Advise from examiners on common pitfalls and how to avoid them

✓ Additional questions that cover the QCA science criteria How Science Works

Please quote S 603 GSA A when ordering

(t) 01865 888068   (f) 01865 314029   (e) orders@heinemann.co.uk   (w) www.heinemann.co.uk

Heinemann
Inspiring generations

L553